IN TIME OF EMERGENCY

A Citizen's Handbook on Nuclear Attack, Natural Disasters

Prepper Press
Your Survival Library

Title: In Time of Emergency
A Citizen's Handbook on Nuclear Attack, Natural Disasters (1968)
Author: Department of Defense
ISBN: 978-1-943544-25-7

This handbook is the property of:

Name______________________________________

Address____________________________________

Location of designated fallout shelter, or shelter nearest to:

Home______________________________________

School_____________________________________

Workplace__________________________________

Emergency telephone numbers:*

Ambulance__________________________________

Civil

Defense____________________________________

Doctors____________________________________

____

Fire_______________________________________

Health

Department_________________________________

Hospitals___________________________________

____

Police_____________________________________

Red

Cross______________________________________

Utility Companies___________________________

____

Weather

Bureau_____________________________________

Other______________________________________

* In a time of nuclear attack or major natural disaster, don't use the telephone to get information or advice. Depend on radio or television.

IN TIME OF EMERGENCY

A Citizen's Handbook on Nuclear Attack, Natural Disasters

The Office of Civil Defense gratefully acknowledges the assistance provided by representatives of the following agencies and organizations in the preparation of material for this handbook:

U.S. Atomic Energy Commission
U.S. Department of Agriculture
U.S. Department of Commerce; Environmental Science Services Administration; Weather Bureau
U.S. Department of Health, Education, and Welfare; Public Health Service
Office of Emergency Planning, Executive Office of the President
American Medical Association; Committee on Disaster Medical Care
American National Red Cross
National Geographic Society
National Association of State Civil Defense Directors
United States Civil Defense Council

The Office of Civil Defense, however, is solely responsible for the validity and accuracy of the information in the handbook.

DEPARTMENT OF DEFENSE
H-14

OFFICE OF CIVIL DEFENSE
March 1968

Table of Contents

Introduction

A major emergency affecting a large number of people may occur anytime and anywhere.

It may be a peacetime disaster such as a flood, tornado, fire, hurricane, blizzard or earthquake. It could be an enemy nuclear attack on the United States.

In any type of general disaster, lives can be saved if people are prepared for the emergency, and know what actions to take when it occurs.

With the aid of Federal and State governments, cities and counties in all parts of the country are developing their local civil defense systems—the fallout shelters, supporting equipment and emergency plans needed to reduce the loss of life from an enemy attack.

While these local government systems have been set up mainly as safeguards against nuclear attack, they have saved lives and relieved suffering in many major peacetime disasters. People have been warned of impending storms and similar dangers, told how to protect themselves, sheltered from the elements, fed and clothed, treated for injury and illness, and given help in resuming their normal lives. Experience has shown that as cities, counties and towns develop their systems to preserve life under nuclear attack conditions, they also become better prepared to deal effectively with peacetime disasters.

In cooperation with the U.S. Office of Civil Defense and the States, many local governments are improving their civil defense systems by preparing community shelter plans. These plans include instructions to local citizens on what to do in the event of nuclear attack.

This handbook, "In Time of Emergency," contains basic general information on both nuclear attack and major natural disasters. This general guidance supplements the specific instructions issued by local governments. Since special conditions may exist in some communities, the local instructions may be slightly different from this general guidance. In those cases, the local instructions should be followed.

Part I is concerned with nuclear attack and basic actions to take.

Part II discusses preparations and emergency actions that will help individuals cope with major natural disasters—floods, hurricanes, tornadoes, winter storms, and earthquakes.

In addition to following the advice given in this handbook and the instructions of their local governments, people can prepare themselves better to meet any major disaster by taking training courses to develop their "emergency skills." Especially recommended are these courses:

"PERSONAL AND FAMILY SURVIVAL" (12-hour course)—A basic orientation course in civil defense, which also tells people how to improve their protection against the effects of a nuclear attack.

"MEDICAL SELF-HELP" (16-hour course)—How to care for the sick and injured if a doctor or nurse is not available.

"FIRST AID" (courses of various lengths)—How to help the sick and injured until professional medical assistance is obtained.

"CARE OF THE SICK AND INJURED" (12-hour course)—How to care for patients after they have received professional medical treatment.

Information on these free courses, which are given in most communities, is available from local Civil Defense Offices, County Agricultural Extension Agents, local public health departments, or American Red Cross chapters. Special advice for rural families on emergency actions related to crops and livestock is available from the U.S. Department of Agriculture.

Part One
NUCLEAR ATTACK

A nuclear attack against the United States would take a high toll of lives. But our losses would be much less if people were prepared to meet the emergency, knew what actions to take, and took them.

A nationwide civil defense system now exists in the United States, and is being enlarged and improved constantly. The heart of this system is fallout shelter to protect people from the radioactive fallout that would result from a nuclear attack. The system also includes warning and communications networks, preparations to measure fallout radiation, control centers to direct lifesaving and recovery operations, emergency broadcasting stations, local governments organized for emergency operations, large numbers of citizens trained in emergency skills, and U.S. military forces available to help civil authorities and the public in a time of emergency.

If an enemy should threaten to attack the United States, you would not be alone. The entire Nation would be mobilizing to repulse the attack, destroy the enemy, and hold down our own loss of life. Much assistance would be available to you—from local, State and Federal governments, from the U.S. armed forces units in your area, and from your neighbors and fellow-Americans. If an attack should come, many lives would be saved through effective emergency preparations and actions.

You can give yourself and your family a much better chance of surviving and recovering from a nuclear attack if you will take time now to:

Understand the dangers you would face in an attack.

Make your own preparations for an attack.

Learn what actions you should take at the time of attack.

Chapter 1
CHECKLIST OF EMERGENCY ACTIONS

✓ **KNOW YOUR LOCAL EMERGENCY ACTION PLAN**

- ✓ Find out from your local government your local plan for emergency action.
- ✓ Determine the specific actions you and members of your family are expected to take.

✓ **UNDERSTAND NUCLEAR ATTACK HAZARDS** (See Chapter 2)

On the widespread threat of fallout, remember:

- ✓ The most dangerous period is the first 24 hours after fallout arrives. But you might have to use fallout shelter for up to two weeks.
- ✓ Highly dangerous amounts of fallout are visible. They look like particles of sand or salt.
- ✓ There is little danger that adults could inhale or swallow enough fallout particles to hurt them. Small children, however, could be injured by drinking contaminated water or milk.
- ✓ A person exposed to fallout radiation does *not* become radioactive. Radiation sickness is *not* contagious; one person cannot "catch it" from another person.

✓ **KNOW THE ATTACK WARNING SIGNAL** (See Chapter 3)

- ✓ On outdoor warning devices, the Attack Warning Signal is a *3- to 5-minute* wavering sound, or a series of short blasts on whistles or horns.

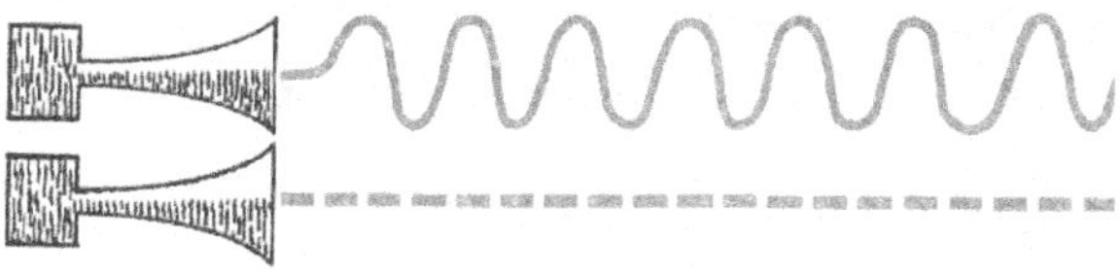

- ✓ This signal means: An enemy attack against the United States has been detected. *Take protective action.* (This signal has no other meaning, and will be used for no other purpose.)
- ✓ On warning, don't use the phone. Get information from radio.

✓ **KNOW THE LOCATION OF FALLOUT SHELTER** (See Chapter 4)

Public shelters are marked like this.

- ✓ Good shelters can be prepared in homes with basements.

✓ **IF NO SHELTER IS AVAILABLE, IMPROVISE PROTECTION** (See Chapter 5)

Remember:

- ✓ A basement corner below ground level, or a storm cellar, is the best place to improvise fallout protection.
- ✓ For the best possible protection, use heavy and dense materials for shielding.

✓ **PREPARE EMERGENCY SUPPLIES** (See Chapter 6)

Especially important are:

- ✓ Water and other liquids.
- ✓ Food requiring no cooking.
- ✓ Special medicines.

✓ **CONSERVE EMERGENCY SUPPLIES; MAINTAIN SANITATION** (See Chapter 7)

✓ **REDUCE FIRE HAZARDS** (See Chapter 8)

✓ **KNOW THE BASICS OF EMERGENCY MEDICAL CARE** (See Chapter 9)

If no doctor is available, especially important are actions to:

- ✓ Restore breathing.

- ✓ Stop serious bleeding.
- ✓ Treat for shock.
- ✓ Treat broken bones and burns.

✓ **FOLLOW OFFICIAL INSTRUCTIONS**

Chapter 2
UNDERSTAND THE HAZARDS OF NUCLEAR ATTACK

Summary

1. The main hazards of a nuclear attack are blast, heat, fire, and fallout radiation.
2. You *may* be able to protect yourself against blast and heat by getting inside a shelter or taking cover, before the nuclear explosions occur. You may be able to avoid fire injuries by putting out small fires or escaping from large fires that might occur in your area.
3. You *can* protect yourself against fallout radiation by getting inside a fallout shelter—if possible, before fallout particles begin drifting down—and by staying there until you are told to come out by authorities who have the equipment to measure radiation levels.
4. After a nuclear attack, food and water would be available to most people, and it would be usable. If any fallout particles have collected, they could be removed before the food is eaten or the water is drunk. People suffering from extreme hunger or thirst should not be denied food or water, even if the available supplies are not known to be free of fallout particles or other radioactive substances.
5. Infants and small children should be fed canned or powdered milk (if available) for awhile after the attack, unless the regular milk supply is uncontaminated. They should not be given water that may contain radioactive substances, if other water known to be pure is available.
6. A person cannot "catch" radiation sickness from another person.

Understand the Hazards of Nuclear Attack

When a nuclear bomb or missile explodes, the main effects produced are intense light (flash), heat, blast, and radiation. How strong these effects are depends on the size and type of the weapon; how far away the explosion is; the weather conditions (sunny or rainy, windy or still); the terrain (whether the ground is flat or hilly); and the height of the explosion (high in the air, or near the ground).

All nuclear explosions cause light, heat and blast, which occur immediately. In addition, explosions that are on or close to the ground would create large quantities of dangerous radioactive fallout particles, most of which would fall to earth during the first 24 hours. Explosions high in the air would create smaller radioactive particles, which would not have any real effect on humans until many months or years later, if at all.*

What Would Happen in an Enemy Attack

If the U.S. should be attacked, the people who happened to be close to a nuclear, explosion—in the area of heavy destruction—probably would be killed or seriously injured by the blast, or by the heat of the nuclear fireball.

---

* These smaller particles would drift to earth more slowly, losing much of their radioactivity before they reached the ground, and would be spread by the upper winds over vast areas of the world.

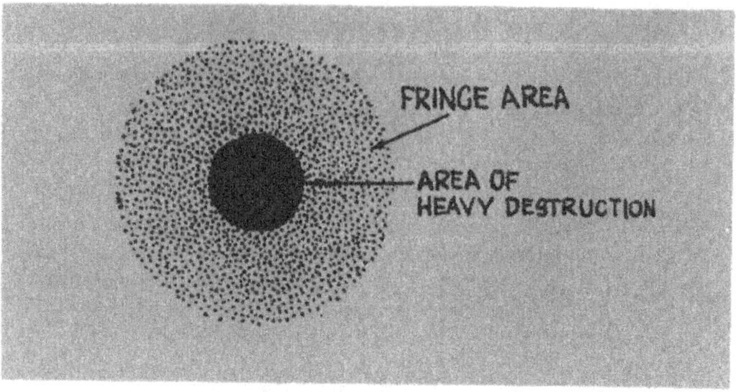

People a few miles away—in the "fringe area" of the explosion—would be endangered by the blast and heat, and by fires that the explosion might start. However, it is likely that most of the people in the fringe area would survive these hazards.

People who were *outside* the fringe area would not be affected by the blast, heat or fire. Department of Defense studies show that in any nuclear attack an enemy might launch against us, tens of millions of Americans would be outside the fringe areas. To them—and to people in the fringe areas who survived the blast, heat and fire—radioactive fallout would be the main danger. Protective measures against this danger can be taken.

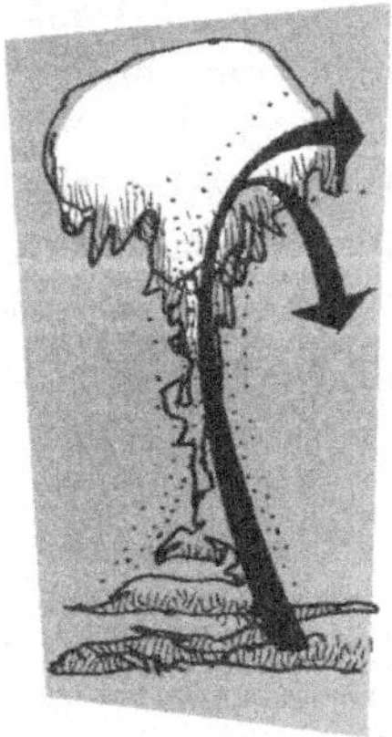

What Is Fallout?

When a nuclear weapon explodes near the ground, great quantities of pulverized earth and other debris are sucked up into the nuclear cloud. There the radioactive gases produced by the explosion condense on and into this debris, producing radioactive fallout particles. Within a short time, these particles fall back to earth—the larger ones first, the smaller ones later. On the way down, and after they reach the ground, the radioactive particles give off invisible gamma rays—like X-rays—too much of which can kill or injure people. These particles give off most of their radiation quickly; therefore the first few hours or days after an attack would be the most dangerous period.

In dangerously affected areas the particles themselves would look like grains of salt or sand; but the *rays* they would give off could not be seen, tasted, smelled or felt. Special instruments would be required to detect the rays and measure their intensity.

Fallout Would Be Widespread

The distribution of fallout particles after a nuclear attack would depend on wind currents, weather conditions and other factors. There is no way of predicting in advance what areas of the country would be affected by fallout, or how soon the particles would fall back to earth at a particular location.

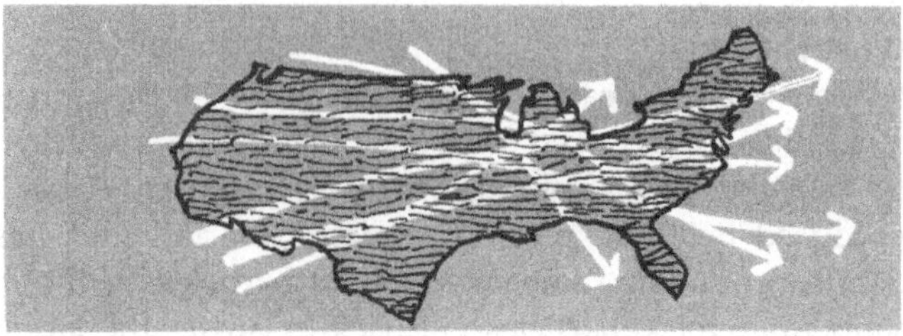

Some communities might get a heavy accumulation of fallout, while others—even in the same general area—might get little or none. No area in the U.S. could be sure of *not* getting fallout, and it is probable that some fallout particles would be deposited on most of the country.

Areas close to a nuclear explosion might receive fallout within 15-30 minutes. It might take 5-10 hours or more for the particles to drift down on a community 100 or 200 miles away.

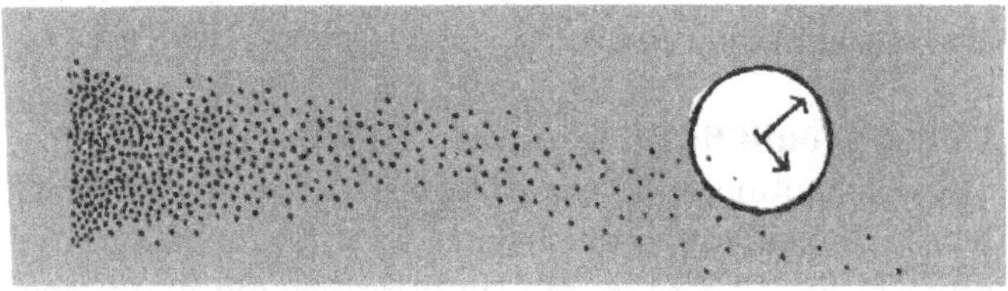

Generally, the first 24 hours after fallout began to settle would be the most dangerous period to a community's residents. The heavier particles falling during that time would still be highly radioactive and give off strong rays. The lighter particles falling later would have lost much of their radiation high in the atmosphere.

Fallout Causes Radiation Sickness

The invisible gamma rays given off by fallout particles can cause radiation sickness—that is, illness caused by physical and chemical changes in the cells of the body. If a person receives a large dose of radiation, he will die. But if he receives only a small or medium dose, his body will repair itself and he will get well. The same dose received over a short period of time is more damaging than if it is received over a longer period. Usually, the effects of a given dose of radiation are more severe in very young and very old persons, and those not in good health.

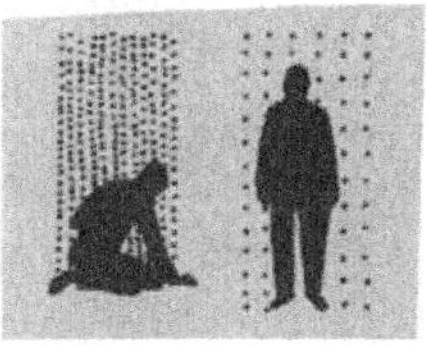

No special clothing can protect people against gamma radiation, and no special drugs or chemicals can prevent large doses of radiation from causing damage to the cells of the body. However, antibiotics and other medicines are helpful in treating infections that sometimes follow excessive exposure to radiation (which weakens the body's ability to fight infections).

Almost all of the radiation that people would absorb from fallout particles would come from particles *outside* their own bodies. Only simple precautions would benecessary to avoid swallowing the particles, and because of their size (like grains of sand) it would be practically impossible to inhale them.

People exposed to fallout radiation do *not* become radioactive and thereby dangerous to other people. Radiation sickness is not contagious or infectious, and one person cannot "catch it" from another person.

Protection Is Possible

People can protect themselves against fallout radiation, and have a good chance of surviving it, by staying inside a fallout shelter. In most cases, the fallout radiation level outside the shelter would decrease rapidly enough to permit people to leave the shelter within a few days.

Even in communities that received heavy accumulations of fallout particles, people soon might be able to leave shelter for a few minutes or a few hours at a time in order to perform emergency tasks. In most places, it is unlikely that full-time shelter occupancy would be required for more than a week or two.

Many Kinds of Fallout Shelters
The farther away you are from the fallout particles outside, the less radiation you will receive. Also, the building materials (concrete, brick, lumber, etc.) that are between you and the fallout particles serve to absorb many of the gamma rays and keep them from reaching you.

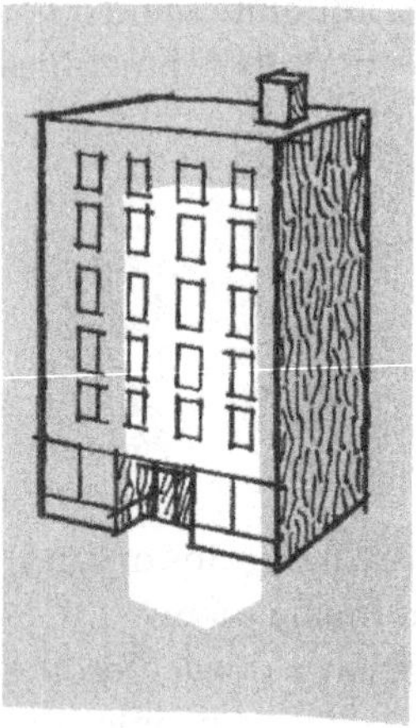

A fallout shelter, therefore, does not need to be a special type of building or an underground bunker. It can be *any space*, provided the walls and roof are thick or heavy enough to absorb many of the rays given off by the fallout particles outside, and thus keep dangerous amounts of radiation from reaching the people inside the structure.

A shelter can be the basement or inner corridor of any large building; the basement of a private home; a subway or tunnel; or even a backyard trench with some kind of shielding material (heavy lumber, earth, bricks, etc.) serving as a roof.

In addition to protecting people from fallout radiation, most fallout shelters also would provide some limited protection against the blast and heat effects of nuclear explosions that were not close by.

Chapter 4 discusses the various types of fallout shelters that people can use to protect themselves in case of nuclear attack.

Food and Water Would Be Available and Usable

From many studies, the Federal Government has determined that enough food and water would be available after an attack to sustain our surviving citizens. However, temporary food shortages might occur in some areas, until food was shipped there from other areas.

Most of the Nation's remaining food supplies would be usable after an attack. Since radiation passing through food does not contaminate it, the only danger would be the actual swallowing of fallout particles that happened to be on the food itself (or on the can or package containing the food), and these could be wiped or washed off. Reaping, threshing, canning and other processing would prevent any dangerous quantities of fallout particles from getting into processed foods. If necessary to further protect the population, special precautions would be taken by food processors.

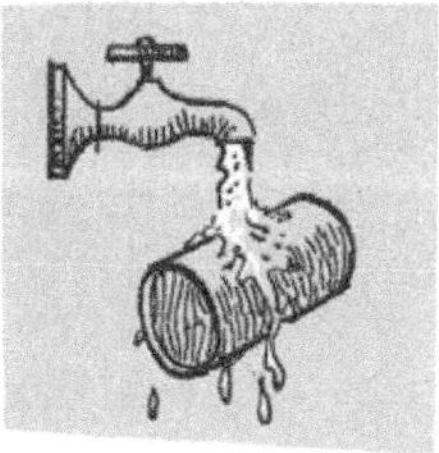

Water systems might be affected somewhat by radioactive fallout, but the risk would be small, especially if a few simple precautions were taken. Water stored in covered containers and water in covered wells would not be contaminated after an attack, because the fallout particles could not get into the water. Even if the containers were not covered (such as buckets or bathtubs filled with emergency supplies of water), as long as they were indoors it is highly unlikely that fallout particles would get into them.

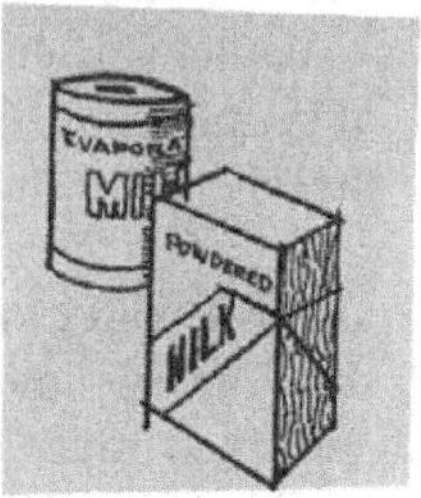

Practically all of the particles that dropped into open reservoirs, lakes, and streams (or into open containers or wells) would settle to the bottom. Any that didn't would be removed when the water was filtered before being pumped to consumers. A small amount of radioactive material might dissolve in the water, but at most this would be of concern for only a few weeks.

Milk contamination from fallout is not expected to be a serious problem after an attack. If cows graze on contaminated pasture and swallow fallout particles that contain some radioactive elements, their milk might be harmful to the thyroid glands of infants and small children. Therefore, if possible, they should be given canned or powdered milk for a few weeks if authorities say the regular milk supply is contaminated by radioactive elements.

In summary, the danger of people receiving harmful doses of fallout radiation through food, water or milk is very small. People suffering from extreme hunger or thirst should not be denied these necessities after an attack, even if the only available supplies might contain fallout particles or other radioactive substances.

Chapter 3
KNOW ABOUT WARNING

Summary

Before an emergency

1. Learn what outdoor warning signals are used in your community, what they sound like, what they mean, and what actions you should take when you hear them.
2. Make sure you know the difference between the Attack Warning Signal and the Attention or Alert Signal (if both are used in your community).

During an emergency

1. When you hear the warning signals, or warning information is broadcast, take prompt action.
2. If the Attack Warning Signal sounds, go to a fallout shelter immediately (unless your local government has told you to do something else). After you are in shelter, listen to a radio for more information and instructions.
3. If there is no public or private shelter you can go to, try to improvise some fallout protection. As a last resort, take cover in the best available place.
4. If there should be a nuclear flash—especially if you feel the warmth from it—take cover instantly, and then move to a fallout shelter later.

Know About Warning

An enemy attack on the United States probably would be preceded by a period of international tension or crisis. This crisis period would help alert all citizens to the *possibility* of attack.

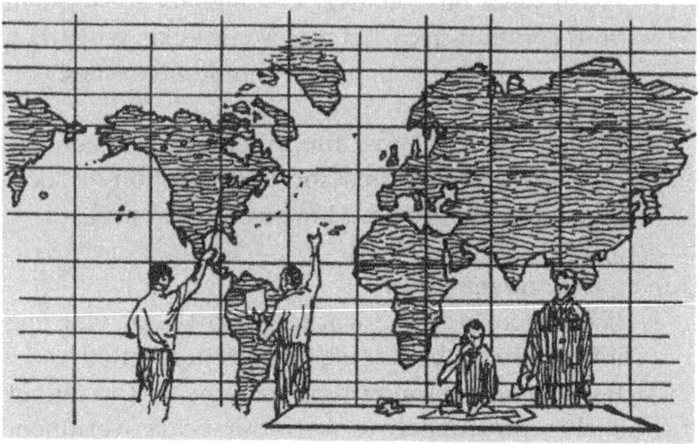

If an attack actually occurs, it is almost certain that incoming enemy planes and missiles would be detected by our networks of warning stations in time for citizens to get into shelters or at least take cover. This warning time might be as little as 5-15 minutes in some locations, or as much as an hour or more in others.

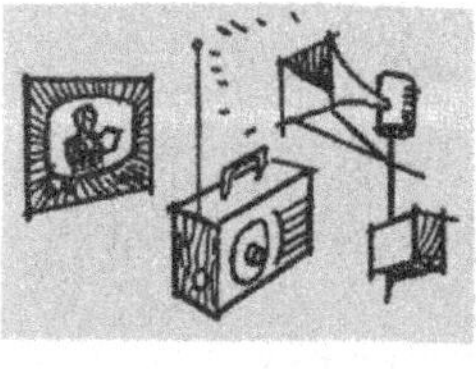

How you received warning of an attack would depend on where you happened to be at that time. You might hear the warning given on radio or television, or even by word-of-mouth. Or your first notice of attack might come from the outdoor warning system in your own city, town or village.

Many U.S. cities and towns have outdoor warning systems, using sirens, whistles, horns or bells. Although they have been installed mainly to warn citizens of enemy attack, some local governments also use them in connection with natural disasters and other peacetime catastrophes.

Different cities and towns are using their outdoor warning systems in different ways. Most local governments, however, have decided to use a certain signal to warn people of an enemy attack, and a different signal to notify them of a peacetime disaster.

The Standard Warning Signals

The two "standard" signals that have been adopted in *most* communities are these:

THE ATTACK WARNING SIGNAL. This will be sounded only in case of enemy attack. The signal itself is a 3- to 5-minute *wavering sound* on the sirens, or a *series of short blasts* on whistles, horns or other devices, repeated as deemed necessary. The Attack Warning Signal means that an actual enemy attack against the United States has been detected, and that protective action should be taken immediately. This signal has no other meaning, and will be used for no other purpose.

THE ATTENTION OR ALERT SIGNAL. This is used by some local governments to get the attention of citizens in a time of threatened or impending natural disaster, or some other peacetime emergency. The signal itself is a 3-to 5-minute *steady blast* on sirens, whistles, horns or other devices. In most places, the Attention or Alert Signal means that the local government wants to broadcast important information on radio or television concerning a peacetime disaster. (See Chapter 1 of Major Natural Disasters section of this handbook.)

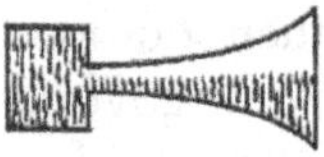

What to Do When Signals Sound

1. *If you should hear the Attack Warning Signal*—unless your local government has instructed you otherwise—go immediately to a public fallout shelter marked like this, or to your home fallout shelter. Turn on a radio, tune it to any local station that is broadcasting, and listen for official information. Follow what-ever instructions are given.

If you are at home and there is no public or private shelter available, you may be able to improvise some last-minute protection for yourself and your family by following the suggestions in Chapter 5 of this handbook. As a last resort, take cover anywhere you can.

2. If you should hear the Attention or Alert Signal, turn on a radio or TV set, tune it to any local station, and follow the official instructions being broadcast.

Don't Use the Telephone

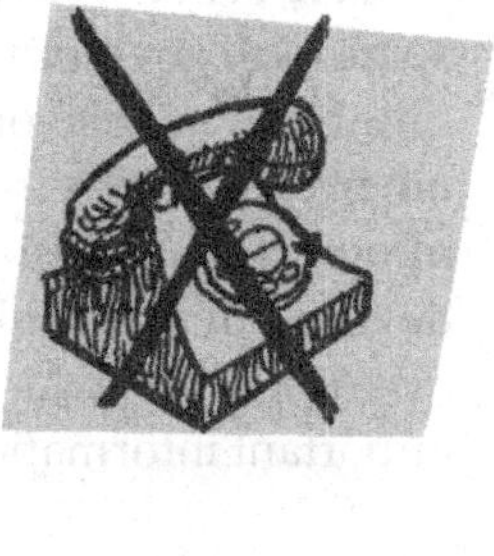

Whichever signal is sounding, *don't* use the telephone to obtain further information and advice about the emergency. Depend on the radio or television, since the government will be broadcasting all the information it has available. The telephone lines will be needed for official calls. Help keep them open.

Learn Your Community's Signals Now

As mentioned before not all communities in the U.S. have outdoor warning systems, and not all communities with warning systems have adopted the two "standard" warning signals.

You should therefore *find out now* from your local Civil Defense Office what signals are being used, in *your community*; what they sound like; what they mean; and what actions you should take when you hear them. Then memorize this information, or write it down on a card to carry with you at all times. Also, post it in your home. Check at least once each year to see if there are any changes.

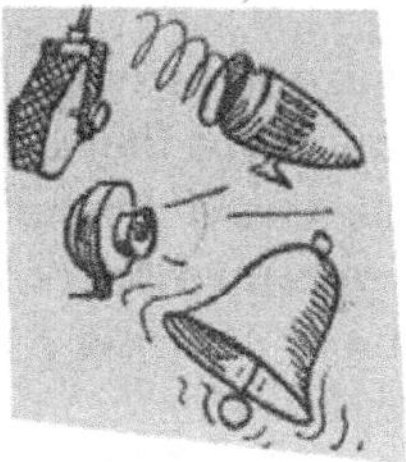

If There Is a Nuclear Flash

It is possible—but extremely unlikely—that your first warning of an enemy attack might be the flash of a nuclear explosion in the sky some distance away. Or there might be a flash after warning had been given, possibly while you were on your way to shelter.

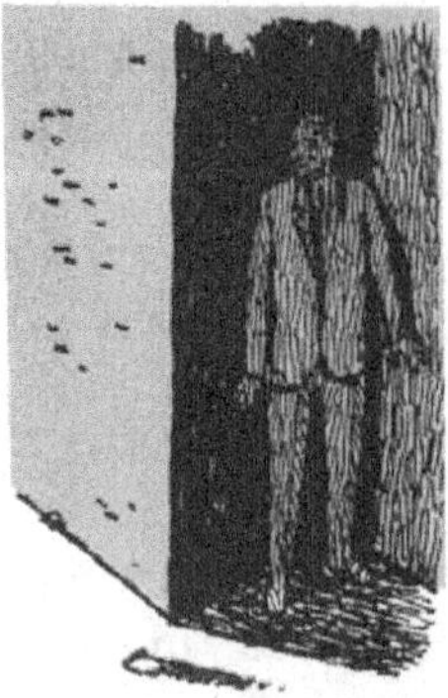

• TAKE COVER INSTANTLY. If there should be a nuclear flash—especially if you are outdoors and feel warmth at the same time—take cover *instantly* in the best place you can find. By getting inside or under something within a few *seconds*, you might avoid being seriously burned by the heat or injured by the blast wave of the nuclear explosion. If the explosion were some distance away, you might have 5 to 15 *seconds* before being seriously injured by the heat, and perhaps 30 to 60 *seconds* before the blast wave arrived. Getting under cover within these time limits might save your life or avoid serious injury. Also, to avoid injuring your eyes, *never look at the flash of an explosion or the nuclear fireball.*

• WHERE TO TAKE COVER. You could take cover in any kind of a building, a storm cellar or fruit cellar, a subway station or tunnel—or even in a ditch or culvert alongside the road, a highway underpass, a storm sewer, a cave or outcropping of rock, a pile of heavy materials, a trench or other excavation. Even getting under a parked automobile, bus or train, or a heavy piece of furniture, would protect you to some extent. If no cover is available, simply lie down on the ground and curl up. The important thing is to avoid being burned by the heat, thrown about by the blast, or struck by flying objects.

• BEST POSITION AFTER TAKING COVER. After taking cover you should lie on your side in a curled-up position, and cover your head with your arms and hands. This would give you some additional protection.

• MOVE TO A FALLOUT SHELTER LATER. If you protected yourself against the blast and heat waves by instantly taking cover, you could get protection from the radioactive fallout (which would arrive later) by moving to a fallout shelter.

Chapter 4
FALLOUT SHELTERS, PUBLIC AND PRIVATE

Summary

Before an emergency

1. Learn the locations of the public fallout shelters that your local government wants you to go to in a time of attack. If no instructions of this kind have been issued, learn the locations of the public shelters nearest to you when you are at home, work, or school. Make sure each member of the family knows these locations.
2. If there is no public fallout shelter near your home, prepare a permanent or preplanned family shelter at home.

During an emergency

1. When you are warned of an enemy attack, go immediately to a public fallout shelter or to your own home shelter, unless your local government has given you other instructions.
2. Stay in shelter until you receive official notice that it is safe to come out.

Fallout Shelters, Public and Private

After a nuclear attack, fallout particles would drift down on most areas of this country. To protect themselves from the radiation given off by these particles, people in affected areas would have to stay in fallout shelters for 2 or 3 days to as long as 2 weeks. Many people would go to public fallout shelters, while others—through choice or necessity—would take refuge in private or home fallout shelters.

Identifying Public Shelters

Most communities now have public fallout shelters that would protect many of their residents against fallout radiation. Where there are still not enough public shelters to accommodate all citizens, efforts are being made to provide more. In the meantime, local governments plan to make use of the best available shelter.

Most of the existing public shelters are located in larger buildings and are marked with this standard yellow-and-black fallout shelter sign. Other public shelters are in smaller buildings, subways, tunnels, mines and other facilities. These also are marked with shelter signs, or would be marked in a time of emergency.

Learn the Locations of Public Shelters

An attack might come at any hour of the day or night. Therefore you should find out *now* the locations of those public fallout shelters designated by your local government for your use. If no designations have yet been made, learn the locations of public shelters that are nearest to you when you are at home, work, school, or any other place where you spend considerable time.

This advice applies to all members of the family. Your children especially should be given clear instructions now on where to find a fallout shelter at all times of the day, and told what other actions they should take in case an attack should occur.

A Home Shelter May Save Your Life

Public fallout shelters usually offer some advantages over home shelters. However, in many places—especially suburban and rural areas—there are few public shelters. If there is none near you, a home fallout shelter may save your life.

The basements of some homes are usable as family fallout shelters as they now stand, without any alterations or changes—especially if the house has two or more stories, and its basement is below ground level.

However, most home basements would need some improvements in order to shield their occupants adequately from the radiation given off by fallout particles. Usually, householders can make these improvements themselves, with moderate effort and at low cost. Millions of homes have been surveyed for the U.S. Office of Civil Defense by the U.S. Census Bureau, and these householders have received information on how much fallout protection their basements would provide, and how to improve this protection.

Shielding Material Is Required

In setting up any home fallout shelter, the basic aim is to place enough "shielding material" between the people in the shelter and the fallout particles outside.

Shielding material is any substance that would absorb and deflect the invisible rays given off by fallout particles outside the house, and thus reduce the amount of radiation reaching the occupants of the shelter. The thicker or denser the shielding material is, the more it would protect the shelter occupants.

Some radiation protection is provided by the existing, standard walls and ceiling of a basement. But if they are not thick or dense enough, other shielding material will have to be added.

Concrete, bricks, earth and sand are some of the materials that are dense or heavy enough to provide fallout protection. For comparative purposes, 4 inches of concrete would provide the same shielding density as:

—5 to 6 inches of bricks.

—6 inches of sand or gravel... }
—7 inches or earth.............. } May be packed into bags, cartons, boxes, or other containers for easier handling.

—8 inches of hollow concrete blocks (6 inches if filled with sand).

—10 inches of water.

—14 inches of books or magazines.

—18 inches of wood.

How to Prepare a Home Shelter

If there is no public fallout shelter near your home, or if you would prefer to use a family-type shelter in a time of attack, you should prepare a home fallout shelter. Here is how to do it:

• A PERMANENT BASEMENT SHELTER. If your home basement—or one corner of it—is below ground level, your best and easiest action would be to prepare a permanent-type family shelter there. The required shielding material would cost perhaps $100-$200, and if you have basic carpentry or masonry skills you probably could do the work yourself in a short time.

Here are three methods of providing a permanent family shelter in the "best" corner of your home basement – that is, the corner which is most below ground level. If you decide to set up one of these shelters, *first get the free plan* for it by writing to Civil Defense, Army Publications Center, 2800 Eastern Blvd. (Middle River), Baltimore, Md. 21220. In ordering a plan, use the full name shown for it.

Ceiling Modification Plan A

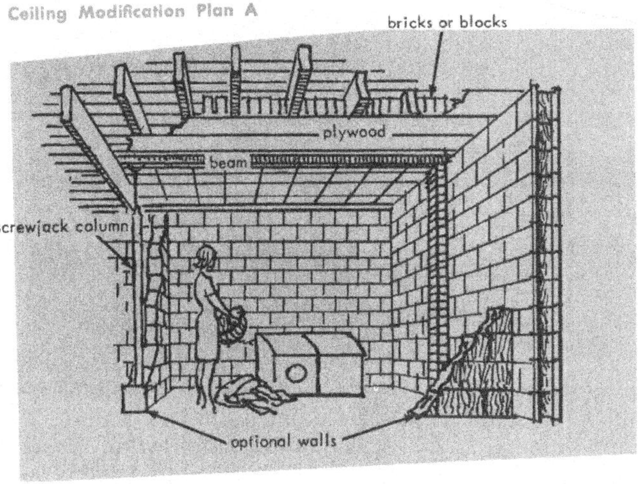

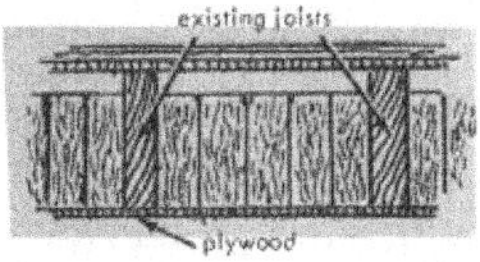

If nearly all your basement is below ground level, you can use this plan to build a fallout shelter area in one corner of it, without changing the appearance of it or interfering with its normal peacetime use.

However, if 12 inches or more of the basement wall is above ground level, this plan should *not* be used unless you add the "optional walls" shown in the sketch.

Overhead protection is obtained by screwing plywood sheets securely to the joists, and then filling the spaces between the joists with bricks or concrete blocks. An extra beam and a screwjack column may be needed to support the extra weight.

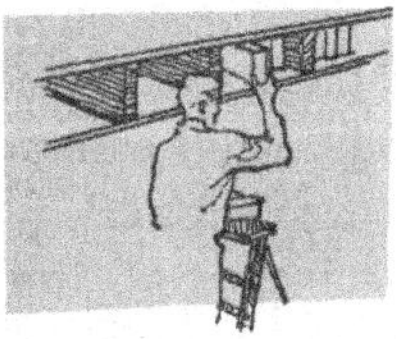

Building this shelter requires some basic woodworking skills and about \$150-\$200 for materials. It can be set up while the house is being built, or afterward.

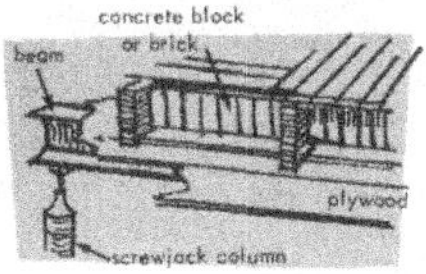

Alternate Ceiling Modification Plan B

This is similar to Plan A, except that new extra joists are fitted into part of the basement ceiling to support the added weight of the shielding (instead of using a beam and a screwjack column).

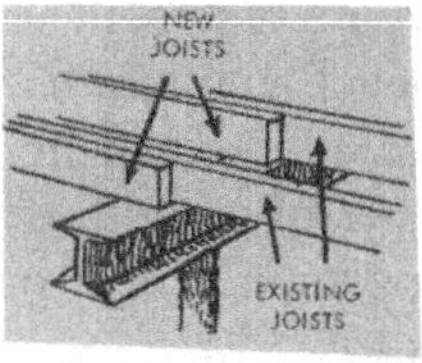

The new wooden joists are cut to length and notched at the ends, then installed between the existing joists.

After plywood panels are screwed securely to the joists, bricks or concrete blocks are then packed tightly into the spaces between the joists. The bricks or blocks, as well as the joists themselves, will reduce the amount of fallout radiation penetrating downward into the basement.

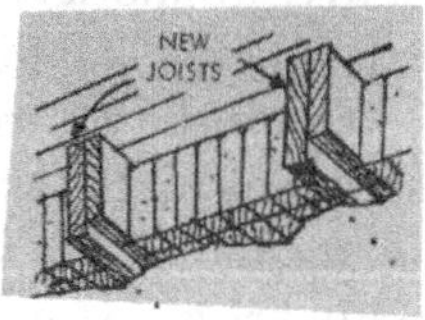

Approximately one-quarter of the total basement ceiling should be reinforced with extra joists and shielding material.

Important: This plan (like Plan A) should *not* be used if 12 inches or more of your basement wall is above ground level, unless you add the "optional walls" inside your basement that are shown in the Plan A sketch.

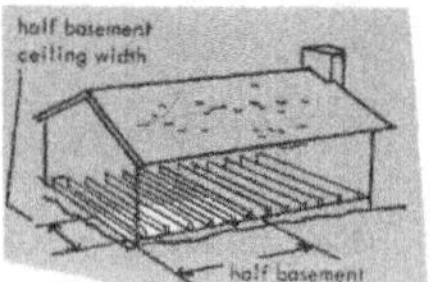

Permanent Concrete Block or Brick Shelter Plan C

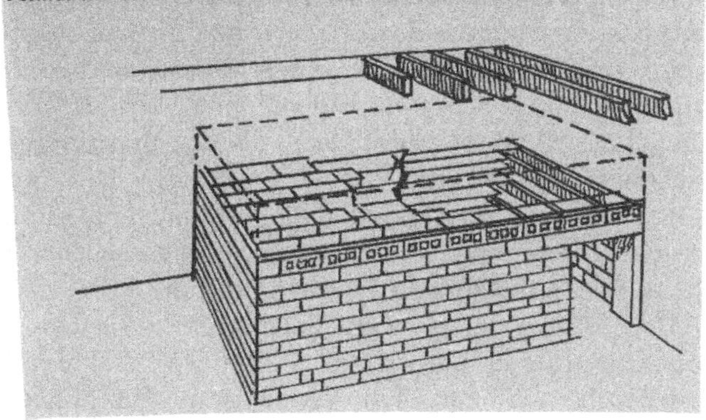

This shelter will provide excellent protection, and can be constructed easily at a cost of $150 in most parts of the country.

Made of concrete blocks or bricks, the shelter should be located in the corner of your basement that is most below ground level. It can be built low, to serve as a "sitdown" shelter; or by making it higher you can have a shelter in which people can stand erect.

The shelter ceiling, however, should *not* be higher than the outside ground level of the basement corner where the shelter is located.

The higher your basement is above ground level, the thicker you should make the walls and roof of this shelter, since your regular basement walls will provide only limited shielding against outside radiation.

Natural ventilation is provided by the shelter entrance, and by the air vents shown in the shelter wall.

This shelter can be used as a storage room or for other useful purposes in non-emergency periods.

A PREPLANNED BASEMENT SHELTER. If your home has a basement but you do not wish to set up a permanent-type basement shelter, the next best thing would be to arrange to assemble a "preplanned" home shelter. This simply means gathering together, in advance, the shielding material you would need to make your basement (or one part of it) resistant to fallout radiation. This material could be stored in or around your home, ready for use whenever you decided to set up your basement shelter.

Here are two kinds of preplanned basement shelters. If you want to set up one of these, be sure to *get the free plan for it first* by writing to Civil Defense, Army Publications Center, 2800 Eastern Blvd. (Middle River), Baltimore, Md. 21220. Mention the full name of the plan you want.

Preplanned Snack Bar Shelter Plan D

This is a snack bar built of bricks or concrete blocks, set in mortar, in the "best" corner of your basement (the corner that is most below ground level). It can be converted quickly into a fallout shelter by lowering a strong, hinged "false ceiling" so that it rests on the snack bar.

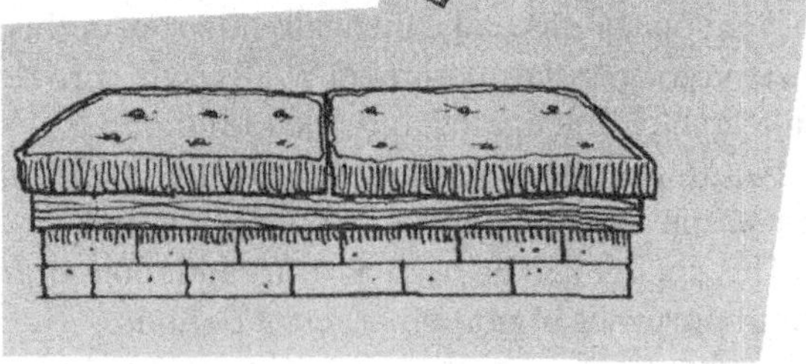

When the false ceiling is lowered into place in a time of emergency, the hollow sections of it can be filled with bricks or concrete blocks. These can be stored conveniently nearby, or can be used as room dividers or recreation room furniture (see bench in sketch).

Preplanned Tilt-Up Storage Unit Plan E

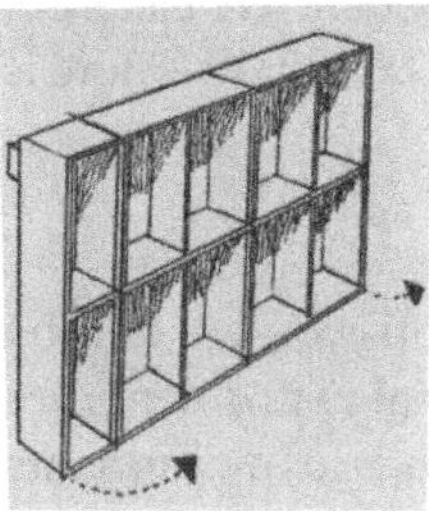

A tilt-up storage unit in the best corner of your basement is another method of setting up a "pre-planned" family fallout shelter.

The top of the storage unit should be hinged to the wall. In peacetime, the unit can be used as a bookcase, pantry, or storage facility.

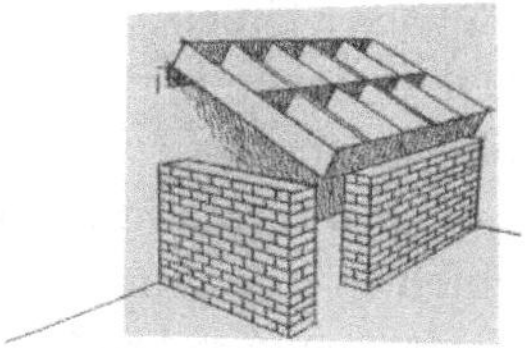

In a time of emergency, the storage unit can be tilted so that the bottom of it rests on a wall of bricks or concrete blocks that you have stored nearby.

Other bricks or blocks should then be placed in the storage unit's compartments, to provide an overhead shield against fallout radiation.

The fallout protection offered by your home basement also can be increased by adding shielding material to the outside, exposed portion of your basement walls, and by covering your basement windows with shielding material.

You can cover the above-ground portion of the basement walls with earth, sand, bricks, concrete blocks, stones from your patio, or other material.

You also can use any of these

substances to block basement windows and thus prevent outside fallout radiation from entering your basement in that manner.

• A PERMANENT OUTSIDE SHELTER. If your home has no basement, or if you prefer to have a permanent-type home shelter in your yard, you can obtain instructions on how to construct several different kinds of outside fallout shelters by writing to the U.S. Office of Civil Defense, Department of Defense, Washington, D.C. 20310. There is no charge for these.

When to Leave Shelter

You should not come out of shelter until you are told by authorities that it is safe to do so. Special instruments are needed to detect fallout radiation and to measure its intensity. Unless you have these instruments, you will have to depend on your local government to tell you when to leave shelter.

This information probably would be given on the radio, which is one reason why you should keep on hand a battery-powered radio that works in your shelter area.

If you came out of shelter too soon, while the fallout particles outside were still highly radioactive, you might receive enough radiation to make you sick or even kill you.

Remember that *fallout particles* can be seen, but the *rays* they give off cannot be seen. If you see unusual quantities of gritty particles outside (on window ledges, sidewalks, cars, etc.) after an attack, you should assume that they are fallout particles, and therefore stay inside your shelter until you are told it is safe to come out.

Chapter 5
IMPROVISING FALLOUT PROTECTION

Summary

Before an emergency

1. If there is no public fallout shelter near your home and you have decided *not* to prepare a permanent or preplanned shelter in your basement or yard, make sure that you have on hand *now* the materials and tools needed to improvise an emergency shelter at home. These would include shielding material (for an inside shelter), and lumber and a shovel (for an outside shelter).

During an emergency

1. If you have no better shelter to go to, improvise an emergency shelter at home.
2. Usually, the best place for an improvised shelter would be in your basement or storm cellar.
3. If you don't have a basement or storm cellar, you might be able to improvise a shelter in the crawl space under your house, outside in your yard, or (as a last resort) on the ground floor of your house. In some places, a boat would provide some fallout protection.

Improvising Fallout Protection

If an enemy attack should occur when you are at home, and you have made no advance shelter preparations, you still might be able to improvise a shelter either inside or outside your house. In a time of emergency, the radio broadcasts may tell you whether you have time to improvise a shelter or should take cover immediately.

An improvised shelter probably would not give you as much protection as a permanent or a preplanned family shelter, but any protection is better than none, and might save your life.

The best place to improvise a shelter would be the basement or storm cellar, if your home has one.

Shielding Material Needed

To improvise a shelter you would need shielding materials such as those mentioned on page 25—concrete blocks, bricks, sand, etc. Other things could also be used as shielding material, or to support shielding material, such as:

—House doors that have been taken off their hinges (especially heavy outside doors).

—Dressers and chests (fill the drawers with sand or earth after they are placed in position, so they won't be too heavy to carry and won't collapse while being carried).

—Trunks, boxes and cartons (fill them with sand or earth after they are placed in position).

—Tables and bookcases.

—Large appliances (such as washers and dryers).

—Books, magazines, and stacks of firewood or lumber.

—Flagstones from outside walks and patios.

Improvising a Basement Shelter

Here are two ways of improvising fallout protection in the basement of a home:

Set up a large, sturdy table or workbench in the corner of your basement that is most below ground level.

On the table, pile as much shielding material as it will hold without collapsing. Around the table, place as much shielding material as possible.

When family members are "inside the shelter"—that is, under the table—block the opening with other shielding material.

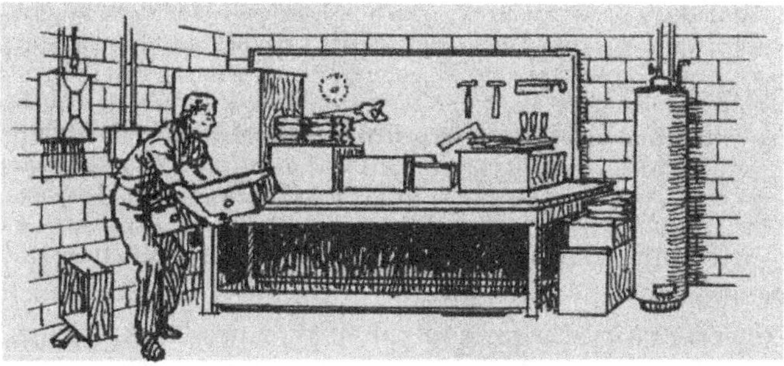

If you don't have a large table or workbench available—or if more shelter space is needed—place furniture or large appliances in the corner of the basement so they will serve as the "walls" of your shelter.

As a "ceiling" for it, use doors from the house that have been taken off their hinges. On top of the doors, pile as much shielding material as they will support. Stack other shielding material around the "walls" of your shelter.

When all persons are inside the shelter space, block the opening with shielding material.

Using a Storm Cellar for Fallout Protection

A below-ground storm cellar can be used as an improvised fallout shelter, but additional shielding material may be needed to provide adequate protection from fallout radiation.

If the existing roof of the storm cellar is made of wood or other light material, it should be covered with one foot of earth or an equivalent thickness of other shielding material for overhead shielding from fallout. More posts or braces may be needed to support the extra weight.

After the roof has been shielded, better protection can be provided by blocking the entrance way with 8-inch concrete blocks or an equivalent thickness of sandbags, bricks, earth or other shielding material, after all occupants are inside the shelter. A few inches should be left open at the top for air. After particles have stopped falling, the outside door may be left open to provide better ventilation.

If shielding material is not available for the entrance way, shelter occupants should stay as far away from it as possible. They also should raise the outside door of the storm cellar now and then to knock off any fallout particles that may have collected on it.

Using the Crawl Space Under Your House

Some homes without basements have "crawl space" between the first floor and the ground underneath the house. If you have this space under your house—and if the house is set on foundation walls, rather than on pillars—you can improvise fallout protection for your family there.

First, get access to the crawl space through the floor or through the outside foundation wall. (A trapdoor or other entry could be made now, before an emergency occurs.)

As the location for your shelter, select a crawl-space area that is under the center of the house, as far away from the outside foundation walls as possible.

Around the selected shelter area, place shielding material—preferably bricks or blocks, or containers filled with sand or earth—from the ground level up to the first floor of the house, so that the shielding material forms the "walls" of your shelter area. On the floor above, place other shielding material to form a "roof" for the shelter area.

If time permits, dig out more earth and make the shelter area deeper, so you can stand erect or at least sit up in it.

Improvising an Outside Shelter

If your home has no basement, no storm cellar and no protected crawl space, here are two ways of improvising fallout protection in your yard:

• Dig an L-shaped trench, about 4 feet deep and 3 feet wide. One side of the L, which will be the shelter area, should be long enough to accommodate all family members. The other side of the L can be shorter, since its purpose is to serve as an entrance-way and to reduce the amount of radiation getting into the shelter area.

Cover the entire trench with lumber (or with house doors that have been taken off their hinges), except for about 2 feet on the short side of the L, to provide access and ventilation.

On top of the lumber or doors, pile earth 1 to 2 feet high, or cover them with other shielding material.

If necessary, support or "shore up" the walls of the trench, as well as the lumber or doors, so they will not collapse.

• Dig a shallow ditch, 6 inches deep and 6 inches wide, parallel to and 4 feet from the outside wall of your house.

Remove the heaviest doors from the house. Place the bottoms of the doors in the ditch (so they won't slip), and lean the doors against the wall of the house.

On the doors, pile 12 to 18 inches of earth or sand. Stack or pile other shielding material at the sides of the doors, and also on the other side of the house wall (to protect you against radiation coming from that direction).

If possible, make the shelter area deeper by digging out more earth inside it. Also dig some other shallow ditches, to allow rain water to drain away.

An Improvised Shelter on the Ground Floor

If your home has no basement or storm cellar (and no crawl space that is surrounded by foundation walls up to the first floor), you can get some limited fallout protection by improvising a fallout shelter on the first or ground floor of your house. However, this type of shelter probably would not give you nearly as much protection as the other types of improvised shelters described in this chapter.

Use an inner hall, inner room or large clothes closet on the ground floor, away from outside walls and windows.

With doors, furniture and appliances, plus stacks of other shielding material, you can create an enclosure large enough to live in for a short time. If possible, use boxes filled with sand or earth as shielding material, and fill drawers and trunks with sand or earth.

If there is not room for the shielding material in the limited space of a closet or small room, you can place the material on the other sides of the walls, or on the floor overhead.

Boats as Improvised Shelters

If no better fallout protection is available, a boat with an enclosed cabin could be used. However, in addition to emergency supplies such as food, drinking water and a battery-powered radio, you should have aboard the items you would need (a broom, bucket, or pump-and-hose) to sweep off or flush off any fallout particles that might collect on the boat.

The boat should be anchored or cruised slowly at least 200 feet offshore, where the water is at least 5 feet deep. This distance from shore would protect you from radioactive fallout particles that had fallen on the nearby land. A 5-foot depth would absorb the radiation from particles falling into the water and settling on the bottom.

If particles drift down on the boat, stay inside the cabin most of the time. Go outside now and then, and sweep or flush off any particles that have collected on the boat.

Chapter 6
SUPPLIES FOR FALLOUT SHELTERS

Summary

Before an emergency

1. If you intend to go to a *public* fallout shelter in a time of attack, find out *now* whether it has emergency supplies in it.
   —If it *has* emergency supplies, always keep on hand at home (or in your car) those few additional supplies you would need to take with you.
   —If it *does not have* emergency supplies, always keep on hand at home all the supplies you would need to take with you.
2. If you intend to use a family fallout shelter at home, always keep on hand, in and around your home, all the supplies and equipment you would need for a shelter stay of two weeks.

During an emergency

1. If you are going to a *public* fallout shelter, take with you the supplies you will need.
2. If you are going to your *home* fallout shelter, gather up the supplies and equipment you want to take to the shelter area with you.

Supplies for Fallout Shelters

People gathered in public and private fallout shelters to escape fallout radiation after a nuclear attack would have to stay there—at least part of the time—for a week or two.

During this time they would need certain supplies and equipment in order to stay alive and well, and to cope with emergency situations that might occur in their shelters.

This chapter tells you what supplies and equipment to take with you if you go to a public fallout shelter, and what items you should keep on hand if you plan to use a family fallout shelter at home.

What to Take to a Public Fallout Shelter

To augment the supply of food and liquids usually found in large buildings, most public fallout shelters are stocked—and others are being stocked—with emergency supplies. These include water containers, emergency food rations, sanitation items, basic medical supplies, and instruments to measure the radiation given off by fallout particles.

If the public shelter you will use in a time of attack contains these or other emergency supplies, you should plan to take with you only these additional items:

—Special medicines or foods required by members of your family, such as insulin, heart tablets, dietetic food or baby food.

—A blanket for each family member.

—A battery-powered radio, a flashlight, and extra batteries.

If the public shelter you are going to does *not* contain emergency supplies, you should take with you all the above items, *plus* as much potable liquids (water, fruit and vegetable juices, etc.) and ready-to-eat food as you can carry to the shelter.

Stocks for a Home Shelter

If you intend to use a home fallout shelter, you should *gather together now* all the things you and your family would need for 2 weeks, even though you probably wouldn't have to remain inside shelter for that entire period.

All these items need not be stocked in your home shelter area. They can be stored elsewhere in or around your house, as long as you could find them easily and move them to your shelter area quickly in a time of emergency.

• **The Absolute Necessities.** There are a few things you must have. They are water, food, sanitation supplies, and any special medicines or foods needed by family members such as insulin, heart tablets, dietetic food and baby food.

• **The Complete List.** In addition to the absolute necessities, there are other important items. Some of them may be needed to save lives. At the least, they will be helpful to you. Here is a list of all major items—both essential and desirable.

WATER. This is even more important than food. Enough water should be available to give each person at least one quart per day for 14 days. Store it in plastic containers, or in bottles or cans. All should have tight stoppers. Part of your water supply might be "trapped" water in the pipes of your home plumbing system, and part of it might be in the form of bottled or canned beverages, fruit or vegetable juices, or milk. A water-purifying agent (either water-purifying tablets, or 2 percent tincture of iodine, or a liquid chlorine household bleach) should also be stored, in case you need to purify any cloudy or "suspicious" water that may contain bacteria.

FOOD. Enough food should be kept on hand to feed all shelter occupants for 14 days, including special foods needed by infants, elderly persons, and those on limited diets. Most people in shelter can get along on about half as much food as usual. If possible store canned or sealed-package foods, preferably those not requiring refrigeration or cooking. These should be replaced periodically. Here is a table showing the suggested replacement periods, in months, for some of the types of food suitable to store for emergency use.*

* This table, and other suggestions concerning emergency supplies of food and water, is contained in "Family Food Stockpile for Survival," Home and Garden Bulletin No. 77, U.S. Department of Agriculture. For sale by the Superintendent of Documents, Washington, D.C. 204022, price 10 cents.

Milk: **Months**

Evaporated..............6

Nonfat dry or whole dry milk, in metal container.................6

Canned meat, poultry, fish: Meat, poultry...18

Fish.......................12

Mixtures of meats, vegetables, cereal products.................18

Condensed meat-and-vegetable Soups....................8

Fruits and vegetables:

Berries and sour cherries, canned.........6

Citrus fruit juices, canned.....................6

Other fruits and fruit juices, canned...........18

Dried fruit, in metal container.................6

Tomatoes, sauerkraut, canned.....................6

Other vegetables, canned (including dry beans and dry peas)..18

Cereals and baked goods: **Months**

Ready-to-eat cereals:

In metal container...12

In original paper package....................1

Uncooked cereal (quick-cooking or instant):

In metal container...24

In original paper package...................12

Hydrogenated (or antioxidant-treated) fats, vegetable oil.............12

Sugars, sweets, nuts:

Sugar......will keep indefinitely

Hard candy, gum.......................18

Nuts, canned.........12

Instant puddings....12

Miscellaneous:

Coffee, tea, cocoa (instant)..................18

Dry cream product (instant)..................12

Bouillon products.................12

Flavored beverage powders.................24

Salt............will keep indefinitely

Flavoring extracts (e.g., pepper)............24

Soda, baking powder .12

SANITATION SUPPLIES. Since you may not be able to use your regular bathroom during a period of emergency, you should keep on hand these sanitation supplies: A metal container with a tight-fitting lid, to use as an emergency toilet; one or two large garbage cans with covers (for human wastes and garbage); plastic bags to line the toilet container; disinfectant; toilet paper; soap; wash cloths and towels; a pail or basin; and sanitary napkins.

MEDICINES AND FIRST AID SUPPLIES. This should include any medicines being regularly taken, or likely to be needed, by family members. First aid supplies should include all those found in a good first aid kit (bandages, antiseptics, etc.), plus all the items normally kept in a well-stocked home medicine chest (aspirin, thermometer, baking soda, petroleum jelly, etc.). A good first aid handbook is also recommended.

INFANT SUPPLIES. Families with babies should keep on hand a two-week stock of infant supplies such as canned milk or baby formula, disposable diapers, bottles and nipples, rubber sheeting, blankets and baby clothing. Because water for washing might be limited, baby clothing and bedding should be stored in larger-than-normal quantities.

COOKING AND EATING UTENSILS. Emergency supplies should include pots, pans, knives, forks, spoons, plates, cups, napkins, paper towels, measuring cup, bottle opener, can opener, and pocket knife. If possible, disposable items should be stored. A heat source also might be helpful, such as an electric hot plate (for use if power is available), or a camp stove or canned-heat stove (in case power is shut off). However, if a stove is used indoors, adequate ventilation is needed.

CLOTHING. Several changes of clean clothing—especially undergarments and socks or stockings—should be ready for shelter use, in case water for washing should be scarce.

BEDDING. Blankets are the most important items of bedding that would be needed in a shelter, but occupants probably would be more comfortable if they also had available pillows, sheets, and air mattresses or sleeping bags.

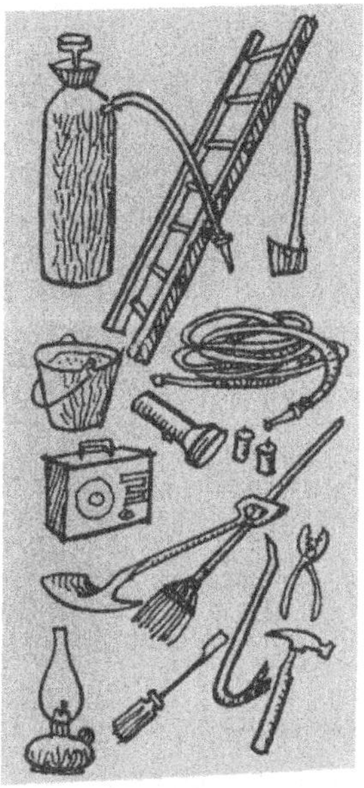

FIRE FIGHTING EQUIPMENT. Simple fire fighting tools, and knowledge of how to use them, may be very useful. A hand-pumped fire extinguisher of the inexpensive, 5-gallon, water type is preferred. Carbon tetrachloride and other vaporizing-liquid type extinguishers are not recommended for use in small enclosed spaces, because of the danger of fumes. Other useful fire equipment for home use includes buckets filled with sand, a ladder, and a garden hose.

GENERAL EQUIPMENT AND TOOLS. The essential items in this category are a battery-powered radio and a flashlight or lantern, with spare batteries. The radio might be your only link with the outside world, and you might have to depend on it for all your information and instructions, especially for advice on when to leave shelter. Other useful items: a shovel, broom, axe, crowbar, kerosene lantern, short rubber hose for siphoning, coil of half-inch rope at least 25 feet long, coil of wire, hammer, pliers, screwdriver, wrench, nails and screws.

MISCELLANEOUS ITEMS. In addition to such practical items as matches, candles, and civil defense instructions, some personal convenience items could be brought into the home shelter if space permits. These might include books and magazines, writing materials, a clock and calendar, playing cards and hobby materials, a sewing kit, and toiletries such as toothbrushes, cosmetics, and shaving supplies.

Chapter 7
WATER, FOOD, AND SANITATION IN A SHELTER

Summary

Before an emergency

1. Read this chapter fully, and learn how you would have to manage your water, food and sanitation problems if you had to spend a week or two in a fallout shelter, especially a home shelter.

During an emergency

1. If you are in a *public* fallout shelter, do exactly what the shelter manager tells you to do. He will take care of you to the best of his ability.
2. If you are in a *home* shelter, follow the advice given in this chapter concerning water, food and sanitation. Take care of your water and food supplies, keep them clean, and make them last for the period you may have to stay in shelter. If necessary, set up an emergency toilet, keep it clean, and make sure it is used properly.

Water, Food, and Sanitation in a Shelter

At all times and under all conditions, human beings must have sufficient water, adequate food and proper sanitation in order to stay alive and healthy. When people are living in a fallout shelter—even for a week or two—water and food may be scarce, and it may be difficult to maintain normal sanitary conditions. Water and food supplies may have to be "managed"—that is, taken care of, kept clean, and rationed to each person in the shelter. Sanitation also may have to be managed and controlled, perhaps by setting up emergency toilets and rules to insure that they are used properly.

If you go to a *public* fallout shelter in a time of attack, you probably would not need to know a great deal about managing water, food, and sanitation. A shelter manager and his assistants would handle these problems with the cooperation of all in the shelter. He would make the best use of whatever water and food supplies were available, provide emergency toilets if necessary, set up rules for living in the shelter, arrange for the shelter occupants to carry on various activities necessary for health and well-being, and decide when it was safe for the group to leave shelter and for how long at a time.

In a *home* fallout shelter, however, you and your family would be largely on your own. You would have to take care of yourselves, solve your own problems, make your own living arrangements, subsist on the supplies you had previously stocked, and find out for yourself (probably by listening to the radio) when it was safe to leave shelter. In this situation, one of your most important tasks would be to manage your water and food supplies, and maintain sanitation. The following guidance is intended to help you do this.

Care and Use of Water Supplies

The average person in a shelter would need at least 1 quart of water or other liquids per day to drink, but more would be useful (to allow some for washing, etc.). Therefore a rationing plan might be required in your home shelter, so as to make your available liquids last for 14 days. (Many communities may continue to have potable water available, and families could relax their rationing plans.)

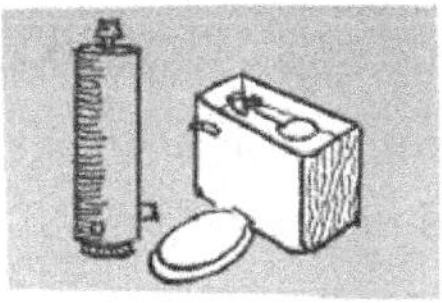

In addition to water stored in containers, there is usually other water available in most homes that is drinkable, such as:

—Water and other liquids normally found in the kitchen, including ice cubes, milk, soft drinks, and fruit and vegetable juices.

—Water (20 to 60 gallons) in the hot water tank.

—Water in the *flush tanks* (not the bowls) of home toilets.

—Water in the pipes of your home plumbing system. In a time of nuclear attack, local authorities may instruct householders to *turn off* the main water valves in their homes to avoid having water drain away in case of a break and loss of pressure in the water mains. With the main valve in your house closed, all the pipes in the house would still be full of water. To use this water, *turn on* the faucet that is located at the *highest* point in your house, to let air into the system; and then draw water, as needed, from the faucet that is located at the lowest point in your house.

In a home shelter, occupants should drink first the water they know is uncontaminated, such as that mentioned above. Of course, if local authorities tell you the regular water is drinkable, it should be used.

If necessary, "suspicious" water—such as cloudy water from regular faucets or perhaps some muddy water from a nearby stream or pond—can be used after it has been purified. This is how to purify it:

1. Strain the water through a paper towel or several thicknesses of clean cloth, to remove dirt and fallout particles, if any. Or else let the water "settle" in a container for 24 hours, by which time any solid particles would have sunk to the bottom. A handful of clay soil in each gallon of water would help this settling process.

2. After the solid particles have been removed, boil the water if possible for 3 to 5 minutes, or add a water-purifying agent to it. This could be either: *(a)* water-purifying tablets, available at drug stores, or *(b)* two percent tincture of iodine, or *(c)* liquid chlorine household bleach, provided the label says that it contains hypochlorite as its *only* active ingredient. For each gallon of water, use 4 water-purifying tablets, or 12 drops of tincture of iodine, or 8 drops of liquid chlorine bleach. If the water is cloudy, these amounts should be doubled.

There would not be much danger of drinking radioactive particles in water, as they would sink quickly to the bottom of the container or stream. Very few would dissolve in the water. Although open reservoirs might contain some radioactive iodine in the first few days after an attack, this danger is considered minor except to very young children.

Care and Use of Food Supplies

Food also should be rationed carefully in a home shelter, to make it last for at least a 2-week period of shelter occupancy. Usually, half the normal intake would be adequate, except for growing children or pregnant women.

In a shelter, it is especially important to be sanitary in the storing, handling and eating of food, so as to avoid digestive upsets or other more serious illness, and to avoid attracting vermin. Be sure to:

–Keep all food in covered containers.

–Keep cooking and eating utensils clean.

–Keep all garbage in a closed container, or dispose of it outside the home when it is safe to go outside. If possible, bury it. Avoid letting garbage or trash accumulate inside the shelter, both for fire and sanitation reasons.

Emergency Toilet Facilities

In many home shelters, people would have to use emergency toilets until it was safe to leave shelter for brief periods of time.

An emergency toilet, consisting of a watertight container with a snug-fitting cover, would be necessary. It could be a garbage container, or a pail or bucket. If the container is small, a larger container, also with a cover, should be available to empty the contents into for later disposal. If possible, both containers should be lined with plastic bags.

This emergency toilet could be fitted with some kind of seat, especially for children or elderly persons. Or it may be possible to remove the seat from a wooden chair, cut a hole in it, and place the container underneath. For privacy, the toilet could be screened from view.

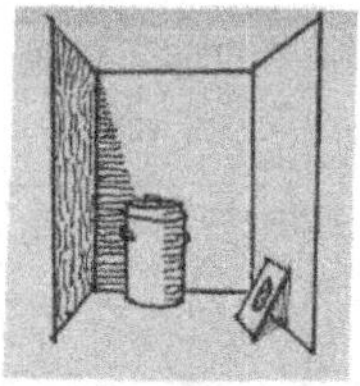

Every time someone uses the toilet, he should pour or sprinkle into it a small amount of regular household disinfectant, such as creosol or chlorine bleach, to keep down odors and germs. After each use, the lid should be put back on.

When the toilet container needs to be emptied, and outside radiation levels permit, the contents should be buried outside in a hole 1 or 2 feet deep. This would prevent the spread of disease by rats and insects.

If the regular toilets inside the home—or the sewer lines—are not usable for any reason, an outside toilet should be built when it is safe to do so.

If anyone has been outside and fallout particles have collected on his shoes or clothing, they should be brushed off before he enters the shelter area again.

Chapter 8
FIRE HAZARDS

Summary

Before an emergency

1. Follow the normal fire prevention rules given in this chapter.
2. Keep on hand at home the basic fire fighting tools mentioned in this chapter.
3. Take a free course in home or auxiliary firefighting, if such a course is offered by the department in your community.

During an emergency

1. Close doors, windows and venetian blinds in your house. Windows not fitted with venetian blinds should be covered with aluminum foil, or coated with whitewash or household cleaning powder, to keep the heat flash of nuclear explosion from entering the house.
2. Unless otherwise advised, fill buckets, bathtubs and other containers with water, for emergency fire-fighting as well as other purposes.
3. If you are in a home fallout shelter, stay there unless you hear an explosion that breaks windows. If you do check to make sure nothing in your home is burning and there are no serious fires nearby.
4. If a fire should occur in your home, fight it promptly, following the recommended procedures in this chapter.

Fire Hazards

Fire, always a danger, could be even more of a disaster during a nuclear attack emergency when the fire department might not be available to help you. Also, the risk of fire would be greater at that time.

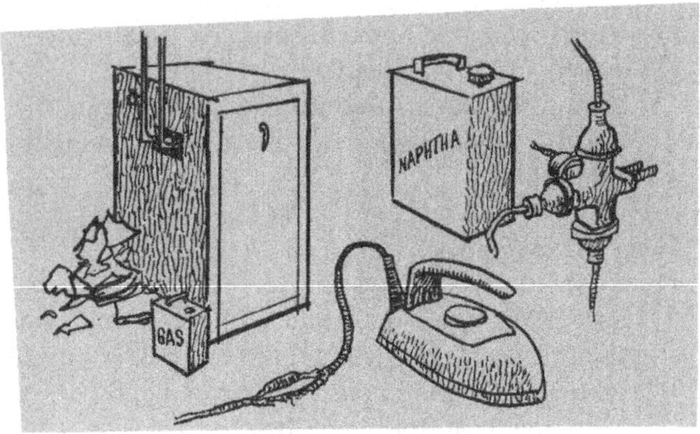

Normal fire-prevention rules are of special importance in an emergency. To keep fires from starting:

Don't let trash and "junk" accumulate. Clean out attics, basements, closets and garages frequently.

Check electrical wiring and appliances. Replace worn or frayed cords. Don't string extension wires over the house, and never under rugs. Use irons and other heating appliances with caution.

Store explosive or flammable fluids carefully, outside the home if possible. Never use gasoline, benzine, naphtha and similar fluids indoors—if their vapor mix with air in a closed space, they will ignite readily from any kind of a spark. Rags soaked with oil or turpentine sometimes catch fire by themselves (this is called spontaneous combustion), and therefore should never be left lying around.

Check heating plants. Many fires are started by faulty furnaces and stoves, cracked or rusted furnace pipes, and sooty chimneys.

Don't place papers or magazines on radiators, or near stoves or fireplaces. Don't allow lamp shades to touch electric bulbs.

These special fire precautions should be taken in a time of nuclear emergency, especially if you plan to use a home shelter:

(1) Keep some of the intense heat rays from nuclear explosions from entering your house by closing your doors, windows, venetian blinds. If windows are not fitted with venetian blinds, cover the inside or outside of the windows with aluminum foil, or coat the glass with whitewash, household cleaning powder, or even mud.

(2) Unless local authorities advise otherwise, fill buckets, bathtubs and other containers with water, for use in emergency fire fighting.

(3) If you have taken refuge in a home fallout shelter, stay there unless you hear an explosion that breaks windows. If you do, check to make sure nothing in your home is burning and there are no other fires nearby that might endanger you.

If a fire does occur, your home might be saved if you know how to fight fires, act promptly, and have on hand some basic firefighting tools. These should include a garden hose (preferably already connected), a ladder, buckets filled with sand, containers filled with water, and a fire extinguisher. Keep in mind that vaporizing-liquid types of fire extinguishers can produce dangerous fumes when used in small enclosed spaces.

Remember the 3 basic ways to put out a fire:

- Take away its fuel.
- Take away its air (smother it).
- Cool it with water or fire-extinguisher chemicals.

Whichever method you use, it will be most effective immediately, before the fire grows larger, by:

—Getting the burning material out of the house (carry it out, or throw it out of a door or window if you can); or

—Putting out the fire with water, sand, earth or fire-extinguisher chemicals; or

—Smothering the fire with a rug or blanket, preferably wet.

Special types of fires require special methods:

—If it is an *electrical fire,* be sure to shut off the electricity first. Then put out the flames with water or anything else available. If you can't shut off the electricity, don't use water on an electrical fire.

—If it is an *oil or grease fire,* shut off the supply of whatever is burning. Then smother the flames with sand, earth, rugs, or other heavy materials. Don't use water.

—If it is a *gas fire,* shut off the gas supply. Then use water, sand, or earth to put out whatever is burning.

The fire departments of some communities in the United States conduct free training courses for citizens interested in learning how to fight fires at home or in becoming auxiliary firemen. If such courses are available in your community, you can acquire fire-fighting skills which may save your live or you home either in peacetime or in a period of nuclear attack.

Chapter 9
EMERGENCY CARE OF THE SICK AND INJURED

Summary

Before an emergency

1. Take the Medical Self-Help course, or a First Aid course.
2. If this is not possible, obtain a good first aid manual, study it, and keep it at home; or study the emergency medical instructions given in this chapter, and keep this handbook at home.
3. Obtain a good first aid kit, and keep your home medicine chest well stocked with supplies you may need in a time of emergency.

During an emergency

1. Try to get a doctor or nurse (or at least a person trained in first aid) to treat anyone who is injured or sick.
2. If no one better qualified is available, take charge yourself.

Emergency Care of the Sick and Injured

A nuclear attack on the United States would cause great numbers of casualties, and there would be fewer doctors, nurses and hospitals available to care for them. Even in areas where no nuclear weapons exploded, radioactive fallout could prevent doctors and nurses from reaching injured or sick persons for a considerable period of time.

People would have to help each other during the emergency. Those in a stocked public fallout shelter would have available the basic medical kit stored there, and perhaps one or more shelter occupants might be a doctor, nurse, or trained first-aider. But persons in a home shelter would have only the medical supplies available at home, and would have to depend on their own knowledge of first aid and emergency medical care.

Both adults and teenagers can acquire these valuable skills now by taking free courses that are offered in many communities, such as the Medical Self-Help course or a First Aid course.

The following information is no substitute for one of these courses. This basic guidance may save lives during a nuclear emergency, however, by helping untrained persons take care of the sick and injured when professional medical assistance may not be immediately available.

General Rules for Any Medical Emergency

1. First of all, *do no harm.* Often, well-meaning but untrained persons worsen the injury or illness in their attempts to help. Get competent medical assistance, if possible. Do not assume responsibility for a patient if you can get the help of a doctor, nurse, or experienced first-aid worker. But if no one better qualified is available, take charge yourself.

2. *Look for stoppage of breathing, and for serious bleeding.* These are the two most life-threatening conditions you can do something about. They demand *immediate* treatment.

3. *Prevent shock, or treat it.* Shock, a serious condition of acute circulatory failure, usually accompanies a severe or painful injury, a serious loss of blood, or a severe emotional upset. If you *expect* shock, and take prompt action, you can prevent it or lessen its severity. This may save the patient's life.

4. *Don't move the patient immediately.* Unless there is real danger of the patient receiving further injury where he is, he should not be moved until breathing is restored, bleeding is stopped, and suspected broken bones are splinted.

5. *Keep calm, and reassure the patient.* Keep him lying down and comfortably warm, but do not apply heat to his body, or make him sweat.

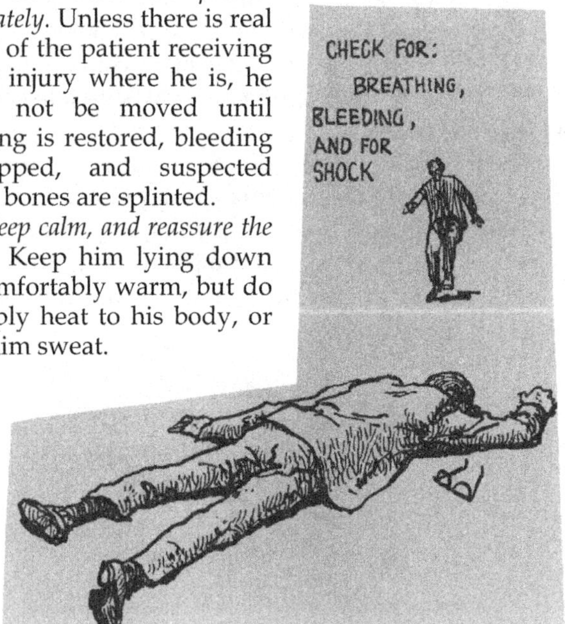

6. *Never attempt to give liquids to an unconscious person.* If he is not able to swallow, he may choke to death or drown. Also, don't give him any liquids to drink if he has an abdominal injury.

If the Patient Has Stopped Breathing

Quick action is required. You must get air into his lungs again immediately or he may die. The best and simplest way of doing this is to use mouth-to-mouth artificial respiration. Here is how to do it:

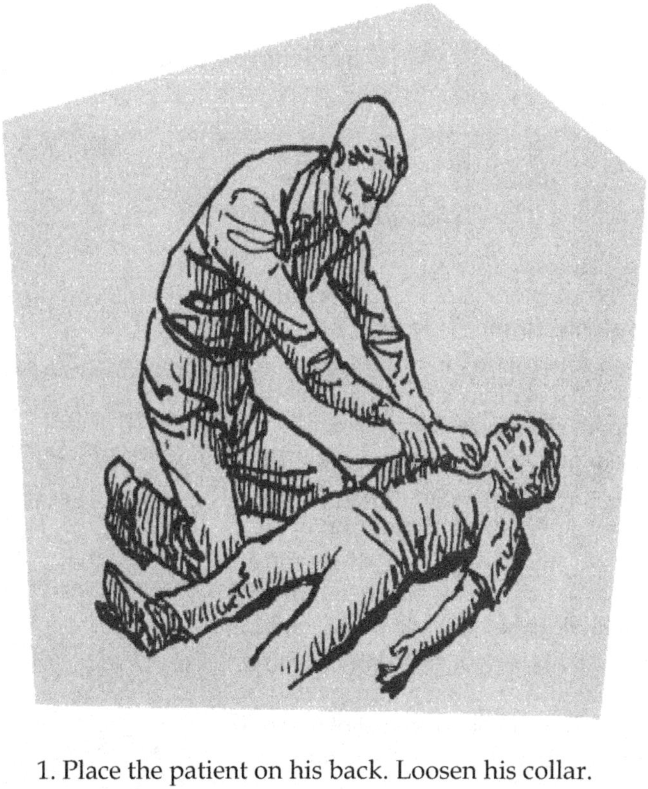

1. Place the patient on his back. Loosen his collar.

2. Open his mouth and use your fingers to remove any food or foreign matter. If he has false teeth or removable

dental bridges, take them out.

3. Tilt the patient's head back so that his chin points upward. Lift his lower jaw from beneath and behind so that it juts out. This will move his tongue away from the back of his throat, so it does not block the air passage to his lungs. Placing a pillow or something else under his shoulders will help get his head into the right position.

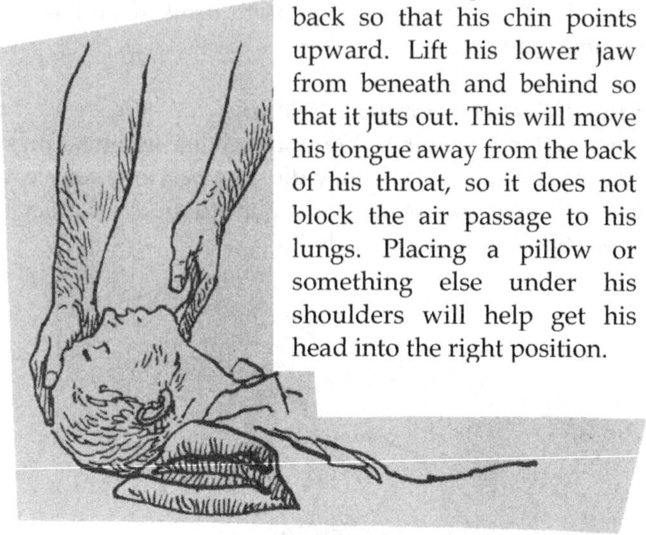

Some patients will start breathing as soon as you take these steps, and no further help is necessary.

4. Open your mouth as wide as possible, and place it tightly over the patient's mouth, so his mouth is completely covered by yours. With one hand, pinch his nostrils shut. With your other hand, hold his lower jaw in a thrust-forward position and keep his head tilted back. With a baby or small child, place your mouth over both his nose and mouth, making a tight seal.

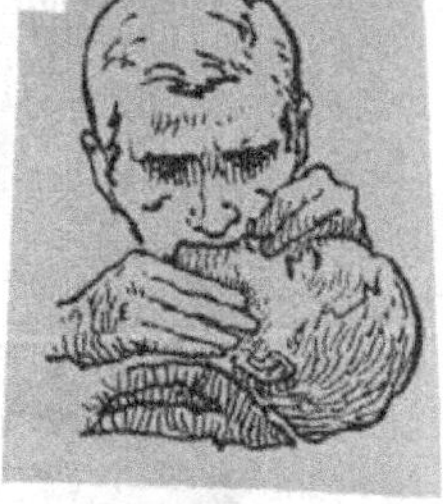

5. Blow a good lungful of air into an adult patient's mouth, continuing to keep his head tilted back and his jaw jutting out so that the air passage is kept open. (Air can be blown

through an unconscious person's teeth, even though they may be clenched tightly together.) Watch his chest as you blow. When you see his chest rise, you will know that you are getting air into his lungs.

6. Remove your mouth from the patient's mouth, and listen for him to breathe out the air you breathed into him. You also may feel his breath on your cheek and see his chest sink as he exhales.

7. Continue your breathing for the patient. If he is an adult, blow a good breath into his mouth every 5 seconds, or 12 times a minute, and listen for him to breathe it back out again. *Caution:* If the patient is an infant or small child, blow *small puffs* of air into him about 20 times a minute. You may rupture his lung if you blow in too much air at one time. Watch his chest rise to make sure you are giving him the right amount of air with each puff.

8. If you are *not* getting air into the patient's lungs, or if he is not breathing out the air you blew into him, first make sure that his head is tilted back and his jaw is jutting out in the proper position. Then use your fingers to make sure nothing in his mouth or throat is obstructing the air passage to his lungs. If this does not help, turn him on his side and strike him sharply with the palm of your hand several times between his shoulder blades. This should dislodge any obstruction in the air passage. Then place him again on his back, with his head tilted back and his jaw jutting out, and resume blowing air into his mouth. If this doesn't work, try closing his mouth and blowing air through his nose into his lungs.

9. If you wish to avoid placing your mouth directly on the patient's face, you may hold a cloth (handkerchief, gauze or other porous material) over his mouth and breathe through the cloth. But don't waste precious time looking for a cloth if you don't have one.

10. *Important*: Even if the patient does not respond, continue your efforts for 1 hour or longer, or until you are completely sure he is dead. If possible, have this confirmed by at least one other person.

To Stop Serious Bleeding

1. Apply firm, even pressure to the wound with a dressing, clean cloth, or sanitary napkin. If you don't have any of these, use your bare hand until you can get something better. Remember, you must keep blood from running out of the patient's body. Loss of 1 or 2 quarts will seriously endanger his life.

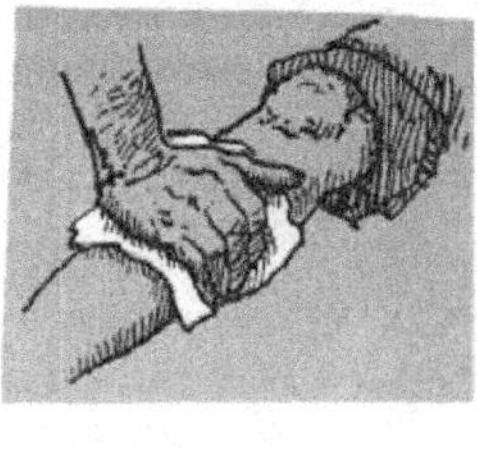

2. Hold the dressing in place with your hand until you can bandage the dressing in place. In case of an arm or leg wound, make sure the bandage is not so tight as to cut off circulation; and raise the arm or leg above the level of the patient's heart. (But if the arm or leg appears broken, be sure to splint it first.)

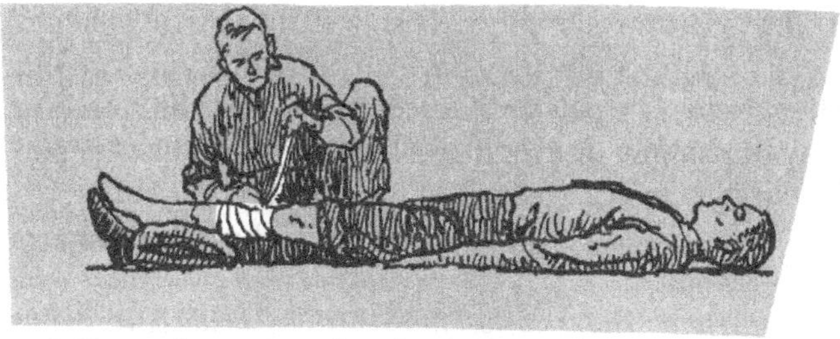

3. Treat the patient for shock.

4. If blood soaks through the dressing, do *not* remove the dressing. Apply more dressings.

5. SPECIAL ADVICE ON TOURNIQUETS: Never use a tourniquet unless you cannot stop excessive, life-threatening bleeding by any other method. Using a tourniquet increases the chances that the arm or leg will have to be amputated later. If you are *forced* to use a tourniquet to keep the patient from bleeding to death (for example, when a hand or foot has been accidentally cut off), follow these instructions carefully:

–Place the tourniquet *as close to the wound as possible,* between the wound and the patient's heart.

–After the tourniquet has been applied, do not permit it to be loosened (even temporarily, or even though the bleeding has stopped) by anyone except a physician, who can control the bleeding by other methods and replace the blood that the patient has lost.

—Get a physician to treat the patient as soon as possible.

Preventing and Treating Shock

Being "in shock" means that a person's circulatory system is not working properly, and not enough blood is getting to the vital centers of his brain and spinal cord.

These are the symptoms of shock: The patient's pulse is weak or rapid, or he may have no pulse that you can find. His skin may be pale or blue, cold, or moist. His breathing may be shallow or irregular. He may have chills. He may be thirsty. He may get sick at his stomach and vomit.

A person can be "in shock" whether he is conscious or unconscious.

Important: All seriously-injured persons should be treated for shock, even though they appear normal and alert. Shock may cause death if not treated promptly, even though the injuries which brought on shock might not be serious enough to cause death. In fact, persons may go into shock without having any physical injuries.

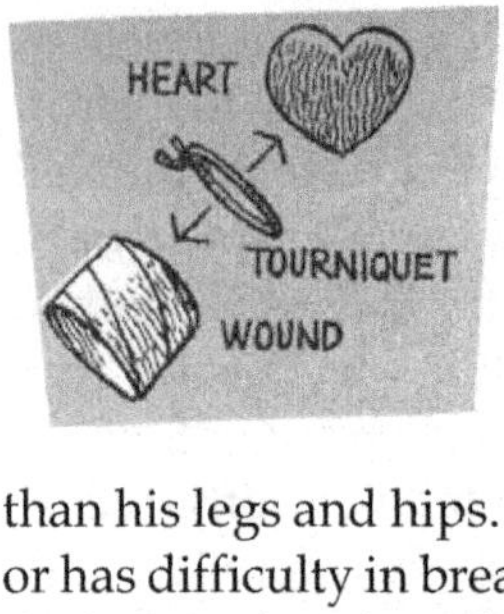

Here is how to treat any person who may be in shock:

1. Keep him lying down and keep him from chilling, but do *not* apply a hot water bottle or other heat to his body. Also, loosen his clothing.

2. Keep his head a little lower than his legs and hips. But if he has a head or chest injury, or has difficulty in breathing, keep his head and shoulders slightly higher than the rest of his body.

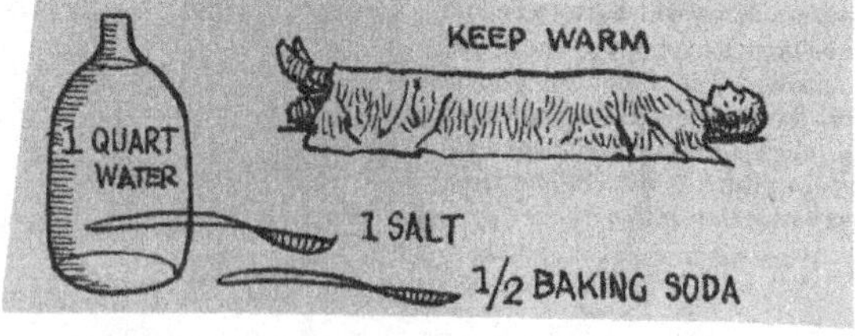

3. Encourage him to drink fluids if he is conscious and not nauseated, and if he does not have abdominal injuries. Every 15 minutes give him a half-glass of this solution until he no longer wants it: One teaspoonful of salt and a half-teaspoonful of baking soda to one quart of water.

4. Do not give him alcohol.

Broken Bones

Any break in a bone is called a fracture. If you think a person may have a fracture, treat it as though it were one. Otherwise, you may cause further injury. For example, if an arm or leg is injured and bleeding, splint it as well as bandage it.

With any fracture, first look for bleeding and control it. Keep the patient comfortably warm and quiet, preferably lying down. If you have an ice bag, apply it to the fracture to ease the pain. Do not move the patient (unless his life is in danger where he is) without first applying a splint or otherwise immobilizing the bone that may be fractured. Treat the patient for shock.

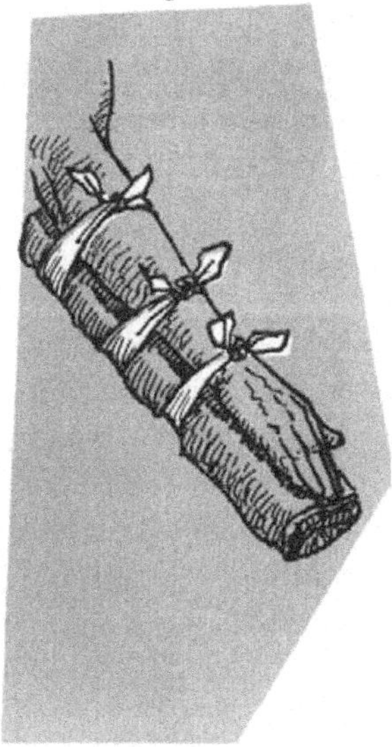

A FRACTURED ARM OR LEG should be straightened out as much as possible, preferably by having 2 persons gently stretch it into a normal position. Then it should be "splinted" — that is, fastened to a board or something else to prevent motion and keep the ends of the broken bone together. As a splint, use a board, a trimmed branch from a tree, a broomstick, an umbrella, a roll of newspapers, or anything else rigid enough to keep the arm or leg straight. Fasten the arm or leg to the

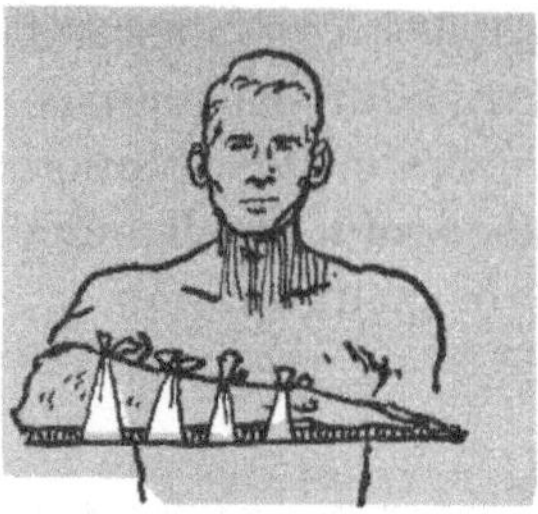

splint with bandages, strips of cloth, handkerchiefs, neckties, or belts. After splinting, keep the injured arm or leg a little higher than the rest of the patient's body. From time to time, make sure that the splint is not too tight, since the arm or leg may swell, and the blood circulation might be shut off. If the broken bone is sticking out through the skin but the exposed part of it is clean, allow it to slip back naturally under the skin (but don't push it in) when the limb is being straightened. However, if the exposed part of the bone is dirty, cover it with a clean cloth and bandage the wound to stop the bleeding. Then splint the arm or leg without trying to straighten it out, and try to find a doctor or nurse to treat the patient.

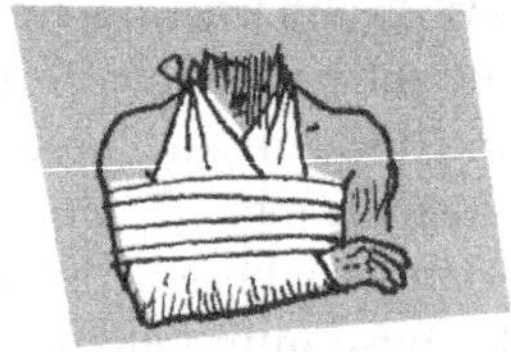

A FRACTURED COLLAR-BONE should also be prevented from moving, until the patient can get professional medical attention. It can be immobilized by placing the arm on that side in a sling and then binding the arm close to the body.

A FRACTURED RIB should be suspected if the patient has received a chest injury or if he has pain when he moves his chest, breathes, or coughs. Strap the injured side of his chest with 2-inch adhesive tape if available, or with a cloth bandage or towel wrapped around and around his entire chest.

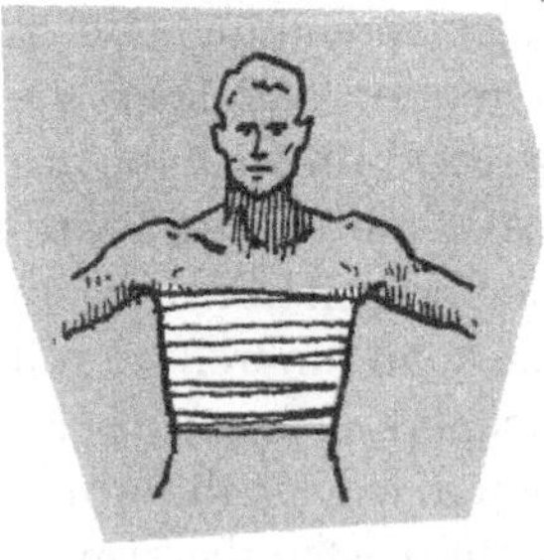

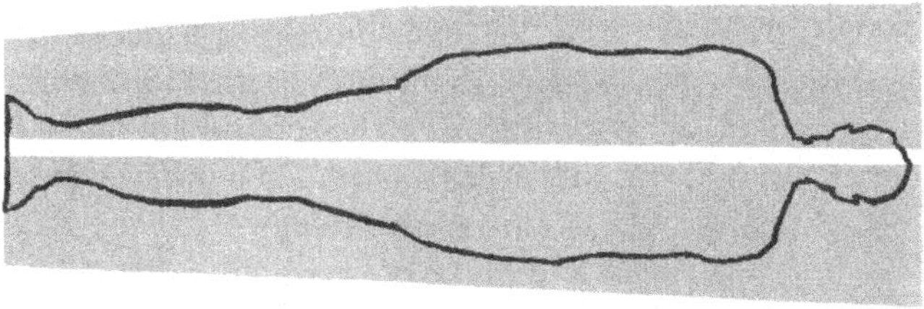

Fractured bones in the NECK OR BACK are very serious, because they may injure the patient's spinal cord and paralyze him or even kill him. He should not be moved until a doctor comes (or a person trained in first aid), unless it is absolutely necessary to move him to prevent further injury. If a person with a back injury has to be moved, he should be placed gently on his back on a stiff board, door or stretcher. His head, back, and legs should be kept in a straight line at all times.

A person with a neck injury should be moved gently with his head, neck, and shoulders kept in the same position they were when he was found. His neck should not be allowed to bend when he is being moved.

Burns

Non-serious or superficial (first degree) burns should not be covered—in fact, nothing need be done for them. However, if a first degree burn covers a large area of the body, the patient should be given fluids to drink as mentioned in item 2 following.

The most important things to do about serious (second or third degree) burns are: *(a)* Treat the patient for shock, *(b)* Prevent infection, and *(c)* Relieve pain. These specific actions should be taken:

1. Keep the patient lying down, with his head a little lower than his legs and hips unless he has a head or chest wound, or has difficulty in breathing.

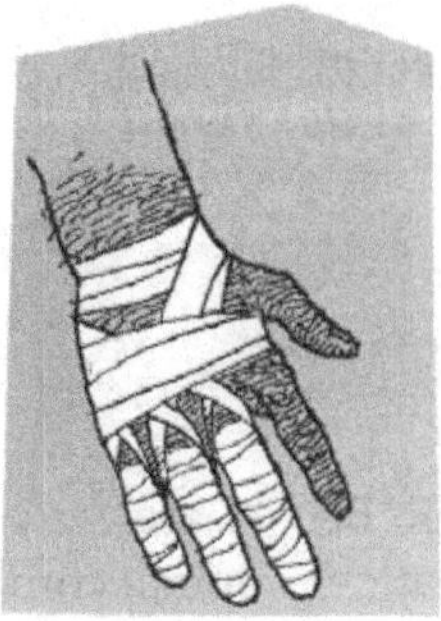

2. Have him drink a half-glass every 15 minutes of a salt-and-soda solution (one teaspoonful of salt and a half-teaspoonful of baking soda to a quart of water). Give him additional plain water to drink if he wants it.

3. Cover the burned area with a *dry,* sterile gauze dressing. If gauze is not available, use a clean cloth, towel or pad.

4. With soap and water, wash the area *around* the burn (not the burn itself) for a distance of several inches, wiping *away* from the burn. The dressing will help prevent surface washings from getting into the burned area.

5. Use a bandage to hold the dry dressing firmly in place against the burned area. This will keep moving air from reaching the burn, and will lessen the pain. Leave dressings and bandage in place as long as possible.

6. If adjoining surfaces of skin are burned, separate them with gauze or cloth to keep them from sticking together (such as between toes or fingers, ears and head, arms and chest).

7. If the burn was caused by a chemical—or by fallout particles sticking to the skin or hair—wash the chemical or the fallout particles away with generous amounts of plain water, then treat the burn as described above.

What NOT to do about burns:

—Don't pull clothing over the burned area (cut it away, if necessary).

—Don't try to remove any pieces of cloth, or bits of dirt or debris, that may be sticking to the burn.

—Don't try to clean the burn; don't use iodine or other

antiseptics on it; and don't open any blisters that may form on it.

– Don't use grease, butter, ointment, salve, petroleum jelly, or any type of medication on severe burns. Keeping them dry is best.
– Don't breathe on a burn, and don't touch it with anything except a sterile or clean dressing.
– Don't change the dressings that were initially applied to the burn, until absolutely necessary. Dressings may be left in place for a week, if necessary.

Radiation Sickness

Radiation sickness is caused by the invisible rays given off by particles of radioactive fallout. If a person has received a large dose of radiation in a short period of time – generally, less than a week – he will become seriously ill and probably will die. But if he has received only a small or medium dose, his body will repair itself and he will get well. No special clothing can protect a person from gamma radiation, and no special medicines can protect him or cure him of radiation sickness.

Symptoms of radiation sickness may not be noticed for several days. The early symptoms are lack of appetite, nausea, vomiting, fatigue, weakness and headache. Later, the patient may have sore mouth, loss of hair, bleeding gums, bleeding under the skin, and diarrhea. But these same symptoms can be caused by other diseases, and not everyone who has radiation sickness shows all these symptoms, or shows them all at once.

If the patient has headache or general discomfort, give him one or two aspirin tablets every 3 or 4 hours (half a tablet, for a child under 12). If he is nauseous, give him "motion sickness tablets," if available. If his mouth is sore or his gums are bleeding, have him use a mouth wash made up of a half-teaspoonful of salt to 1 quart of water. If there is vomiting or diarrhea, he should drink slowly several glasses each day of a salt-and-soda solution (one teaspoonful of salt and one-half teaspoonful of baking soda to 1 quart of cool water), plus bouillon or fruit juices. If available, a mixture of kaolin and pectin should be given for diarrhea. Whatever his symptoms, the patient should be kept lying down, comfortably warm, and resting.

Remember that radiation sickness is not contagious or infectious, and one person cannot "catch it" from another person.

Part Two
MAJOR NATURAL DISASTERS

Many of the actions recommended in Part I of this handbook to help you prepare for and live through a nuclear attack—such as learning the warning signals, stocking emergency supplies, taking a course in emergency skills, and knowing how to fight fires at home—also would help you in case a major natural disaster occurs in your area. If you are prepared for nuclear attack, you are also prepared to cope with most peacetime disasters—disasters that kill hundreds of Americans every year, injure thousands, inflict widespread suffering and hardship, and cause great economic loss.

Part II of this handbook is intended to help you prepare for those natural disasters that may occur in your area, and tell you the right actions to take if they occur. Chapter 1 gives general guidance applicable to various types of natural disasters. Succeeding chapters give special advice on floods, hurricanes, tornadoes, winter storms, and earthquakes.

Chapter 1
GENERAL GUIDANCE

There are certain things you can learn and do that will help you get ready for, and cope with, almost any type of natural disaster.

Perhaps the most basic thing to remember is to *keep calm*. This may mean the difference between life and death. In many disasters, people have been killed or injured needlessly because they took thoughtless actions when they should have done something else — or done nothing at all just then.

In a time of emergency, taking proper action may save your life. *Take time to think*, and then take the considered action that the situation calls for. Usually, this will be the action you have planned in advance, or the action you are instructed to take by responsible authorities.

Here is other guidance that applies to most types of natural disasters.

Warning

LEARN YOUR COMMUNITY'S WARNING SIGNALS. In most communities having outdoor warning systems, the Attack Warning Signal is a wavering sound on the sirens, or a series of short blasts on whistles, horns, or other devices. This signal will be used only to warn of an attack against the United States.

Many communities also are using an *Attention or Alert Signal*, usually a 3- to 5-minute *steady blast* to get the attention of their people in a time of threatened or impending peacetime emergency. In most places, the Attention or Alert Signal means that people should turn on their radio or television sets to hear important emergency information being broadcast.

You should find out now, before any emergency occurs, what warning signals are being used in your community, what they sound like, what they mean, and what actions you should take when you hear them.

Also, whenever a major storm or other peacetime disaster threatens, keep your radio or television set turned on to hear Weather Bureau reports and forecasts (issued by the Environmental Science Services Administration of the U.S. Department of Commerce), as well as other information and advice that may be broadcast by your local government.

When you are warned of an emergency, get your information on the radio or television. Use your telephone only to *report* important events (such as fires, flash floods, or tornado sightings) to the local authorities.

If you tie up the telephone lines simply to get information, you may prevent emergency calls from being completed.

Emergency Supplies

A major disaster of almost any kind may interfere with your normal supplies of water, food, heat, and other day-to-day necessities. You should keep on hand, in or around your home, a stock of emergency supplies sufficient to meet your needs for a few days or preferably for a week.

If you stayed at home during the disaster, these supplies would help you live through the period of emergency without hardship. If you had to evacuate your home and move temporarily to another location, your emergency supplies could be taken with you and used en route or after you arrived at the new location (where regular supplies might not be available). Even if you only had to move to an emergency shelter station set up by a local agency, these supplies might be helpful to you, or make your stay easier.

The most important items to keep on hand are water (preferably in plastic jugs or other stoppered containers); canned or sealed-package foods that do not require refrigeration or heat for cooking; medicines needed by family members, and a first aid kit; blankets or sleeping bags; flashlights or lanterns; a battery-powered radio; and perhaps a covered container to use as an emergency toilet. In addition, an automobile in good operating condition with an ample supply of gasoline may be necessary in case you have to leave your home.

In those parts of the country subject to hurricanes or floods, it is also wise to keep on hand certain emergency materials you may need to protect your home from wind and water—such as plywood sheeting or lumber to board up your windows and doors, and plastic sheeting or tarpaulins to protect furniture and appliances.

Fire Protection and Fire Fighting

Fires are a special hazard in a time of disaster. They may start more readily, and the help of the fire department may not be available quickly. Therefore, it is essential that you:

1. Follow the fire prevention rules given, and be especially careful not to start fires.
2. Know how to put out small fires yourself.
3. Have on hand simple tools and equipment needed for fire fighting.

After a Natural Disaster

Use extreme caution in entering or working in buildings that may have been damaged or weakened by the disaster, as they may collapse without warning. Also, there may be gas leaks or electrical short circuits.

Don't bring lanterns, torches or lighted cigarettes into buildings that have been flooded or otherwise damaged by a natural disaster, since there may be leaking gas lines or flammable material present.

Stay away from fallen or damaged electric wires, which may still be dangerous.

Check for leaking gas pipes in your home. Do this by smell only—don't use matches or candles. If you smell gas, do this: (1) Open all windows and doors, (2) Turn off the main gas valve at the meter, (3) Leave the house immediately, (4) Notify the gas company or the police or fire department, (5) Don't re-enter the house until you are told it is safe to do so.

If any of your electrical appliances are wet, first turn off the main power switch in your house, then unplug the wet appliance, dry it out, reconnect it, and finally, turn on the main power switch. (Caution: Don't do any of these things while *you* are wet or standing in water.) If fuses blow when the electric power is restored, turn off the main power switch again and then inspect for short circuits in your home wiring, appliances and equipment.

Check your food and water supplies before using them. Foods that require refrigeration may be spoiled if electric power has been off for some time. Also, don't eat food that has come in contact with flood waters. Be sure to follow the instructions of local authorities concerning the use of food and water supplies.

If needed, get food, clothing, medical care or shelter at Red Cross stations or from local government authorities.

Stay away from disaster areas. Sightseeing could interfere with first aid or rescue work, and may be dangerous as well.

Don't drive unless necessary, and drive with caution. Watch for hazards to yourself and others, and report them to local authorities.

Write, telegraph or telephone your relatives, after the emergency is over, so they will know you are safe. Otherwise local authorities may waste time locating you—or if you have evacuated to a safer location, they may not be able to find you. (However, do not tie up the phone lines if they are still needed for official emergency calls.)

Do not pass on rumors or exaggerated reports of damage.

Follow the advice and instructions of your local government on ways to help yourself and your community recover from the emergency.

Chapter 2
FLOODS AND HURRICANES

In addition to the general guidance in Chapter 1 of this section, there are certain emergency actions particularly associated with major floods, hurricanes, and storm tides or surges. These types of disasters usually are preceded by extended periods of warning. People living in areas likely to be most severely affected often are warned to move to safer locations.

Evacuation

If you are warned to evacuate your home and move to another location temporarily, there are certain things to remember and do. Here are the most important ones:

• FOLLOW THE INSTRUCTIONS AND ADVICE OF YOUR LOCAL GOVERNMENT. If you are told to evacuate, do so promptly. If you are instructed to move to a certain location, go there—don't go anywhere else. If certain travel routes are specified or recommended, use those routes rather than trying to find short cuts of your own. (It will help if you have previously become familiar with the routes likely to be used.) If you are told to shut off your water, gas or electric service before leaving home, do so. Also find out on the radio where emergency housing and mass feeding stations are located, in case you need to use them.

• SECURE YOUR HOME BEFORE LEAVING. If you have time, and if you have not received other instructions from your local government, you should take the following actions before leaving your home:

—Bring outside possessions inside the house, or tie them down securely. This includes outdoor furniture, garbage cans, garden tools, signs, and other movable objects that might be blown or washed away.

—Board up your windows so they won't be broken by high winds, water, flying objects or debris.

—If flooding is likely, move furniture and other movable objects to the upper floor of your house. Disconnect any electrical appliances or equipment that cannot be moved—but don't touch them if you are wet or are standing in water.

—Do *not* stack sandbags around the outside walls of your house to keep flood waters out of your basement. Water seeping downward through the earth (either beyond the sandbags or over them) may collect around the basement walls and under the floor, creating pressure that could damage the walls or else raise the entire basement and cause it to "float" out of the ground. In most cases it is better to permit the flood waters to flow freely into the basement (or flood the basement yourself with clean water, if you feel sure it will be flooded anyway). This will equalize the water pressure on the inside and outside of the basement walls and floor, and thus avoid structural damage to the foundation and the house.

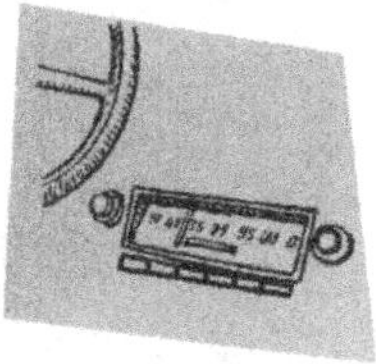

—Lock house doors and windows. Park your car in the garage or driveway, close the windows, and lock it (unless you are driving to your new temporary location).

• TRAVEL WITH CARE. If your local government is arranging transportation for you, precautions will be taken for your safety. But if you are walking or driving your own car to another location, keep in mind these things:

—Leave early enough so as not to be marooned by flooded roads, fallen trees, and wires.

—Make sure you have enough gasoline in your car.

—Follow recommended routes.

—As you travel, keep listening to the radio for additional information and instructions from your local government.

—Watch for washed-out or undermined roadways, earth slides, broken sewer or water mains, loose or downed electric wires, and falling or fallen objects.

—Watch out for areas where rivers or streams may flood suddenly.

—Don't try to cross a stream or a pool of water unless you are certain that the water will not be above your knees (or above the middle of your car's wheels) *all the way across*. Sometimes the water will hide a bridge or a part of the road that has been washed out. If you decide it is safe to drive across it, put your car in low gear and drive very slowly, to avoid splashing water into your engine and causing it to stop. Also, remember that your brakes may not work well after the wheels of your car have been in deep water. Try them out a few times when you reach the other side.

During a Hurricane

—If your house is on high ground and you haven't been instructed to evacuate, stay indoors. Don't try to travel, since you will be in danger from flying debris, flooded roads, and downed wires.

—Keep listening to your radio or television set for further information and advice. If the center or "eye" of the hurricane passes directly over you, there will be a temporary lull in the wind, lasting from a few minutes to perhaps a half-hour or more. *Stay in a safe place during this lull.* The wind will return—perhaps with even greater force—from the *opposite* direction.

Special Advice on Flash Floods

In many areas, unusually heavy rains may cause quick or "flash" floods. Small creeks, gullies, dry streambeds, ravines, culverts or even low-lying grounds frequently flood very quickly and endanger people, sometimes before any warning can be given.

In a period of heavy rains, be aware of this hazard and be prepared to protect yourself against it. If you see any possibility of a flash flood occurring where you are, move immediately to a safer location (don't wait for instructions to move), and then notify your local authorities of the danger, so other people can be warned.

Chapter 3
TORNADOES

• *When a tornado watch (forecast) is announced,* this means that tornadoes are expected in or near your area. Keep your radio or television set tuned to a local station for information and advice from your local government or the Weather Bureau. Also, keep watching the sky, especially to the south and southwest. (When a tornado watch is announced during the approach of a hurricane, however, keep watching the sky to the east.) If you see any revolving, funnel-shaped clouds, report them by telephone immediately to your local police department, sheriff's office or Weather Bureau office. But do not use the phone to get information and advice—depend on radio or TV.

• *When a tornado warning is issued, take shelter immediately.* The warning means that a tornado has actually been sighted, and this (or other tornadoes) may strike in your vicinity. You must take action to protect yourself from being blown away, struck by falling objects, or injured by flying debris. Your best protection is an underground shelter or cave, or a substantial steel-framed or reinforced-concrete building. But if none of these is available, there are other places where you can take refuge:

—If you are at *home*, go to your underground storm cellar or your basement fallout shelter, if you have one. If not, go to a corner of your home basement and take cover under a sturdy workbench or table (but not underneath heavy appliances on the floor above). If your home has no basement, take cover under heavy furniture on the ground floor in the center part of the house, or in a small room on the ground floor that is away from outside walls and windows. (As a last resort, go outside to a nearby ditch, excavation, culvert or ravine.) Doors and windows on the sides of your house *away from*, the tornado may be left open to help reduce damage to the building, but stay away from them to avoid flying debris. Do not remain in a trailer or mobile home if a tornado is approaching; take cover elsewhere.

—If you are at *work* in an office building, go to the basement or to an inner hallway on a lower floor. In a factory, go to a shelter area, or to the basement if there is one.

—If you are *outside in open country*, drive away from the tornado's path, at a right angle to it. If there isn't time to do this—or if you are walking—take cover and lie flat in the nearest depression, such as a ditch, culvert, excavation, or ravine.

Chapter 4
WINTER STORMS

Here is advice that will help you protect yourself and your family against the hazards of winter storms—blizzards, heavy snows, ice storms, freezing rain, or sleet.

• KEEP POSTED ON WEATHER CONDITIONS. Use your radio, television and newspapers to keep informed of current weather conditions and forecasts in your area. Even a few hours' warning of a storm may enable you to avoid being caught outside in it, or at least be better prepared to cope with it. You should also understand the terms commonly used in weather forecasts:

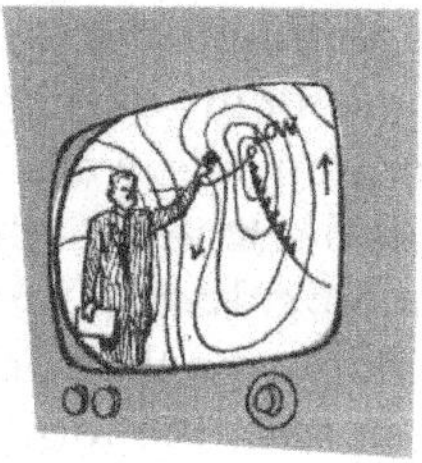

—A *blizzard* is the most dangerous of all winter storms. It combines cold air, heavy snow, and strong winds that blow the snow about and may reduce visibility to only a few yards. A *blizzard warning* is issued when the Weather Bureau expects considerable snow, winds of 35 miles an hour or more, and temperatures of 20 degrees Fahrenheit or lower. A *severe blizzard warning* means that a very heavy snowfall is expected, with winds of at least 45 miles an hour and temperatures of 10 degrees or lower.

—A *heavy snow warning* usually means an expected snowfall of 4 inches or more in a 12-hour period, or 6 inches or more in a 24-hour period. Warnings of *snow flurries, snow squalls, or blowing and drifting snow* are important mainly because visibility may be reduced and roads may become slippery or blocked.

—*Freezing rain or freezing drizzle* is forecast when expected rain is likely to freeze as soon as it strikes the ground, putting a coating of ice or glaze on roads and everything else that is exposed. If a substantial layer of ice is expected to accumulate from the freezing rain, an *ice storm* is forecast.

—*Sleet* is small particles of ice, usually mixed with rain. If enough sleet accumulates on the ground, it will make the roads slippery.

• BE PREPARED FOR ISOLATION AT HOME. If you live in a rural area, make sure you could survive at home for a week or two in case a storm isolated you and made it impossible for you to leave. You should:

—Keep an adequate supply of heating fuel on hand and use it sparingly, as your regular supplies may be curtailed by storm conditions. If necessary, conserve fuel by keeping the house cooler than usual, or by "closing off" some rooms temporarily. Also, have available some kind of *emergency* heating equipment and fuel so you could keep at least one room of your house warm enough to be livable. This could be a camp stove with fuel, or a supply of wood or coal if you have a fireplace. If your furnace is controlled by a thermostat and your electricity is cut off by a storm, the furnace probably would not operate and you would need emergency heat.

—Stock an emergency supply of food and water, as well as emergency cooking equipment such as a camp stove. Some of this food should be of the type that does not require refrigeration or cooking.

—Make sure you have a battery-powered radio and extra batteries on hand, so that if your electric power is cut off you could still hear weather forecasts, information and advice broadcast by local authorities. Also, flashlights or lanterns would be needed.

—Consult this handbook for other supplies and equipment that you may need if isolated at home. Be sure to keep on hand the simple tools and equipment needed to fight a fire. Also, be certain that all family members know how to take precautions that would prevent fire at such a time, when the help of the fire department may not be available.

• TRAVEL ONLY IF NECESSARY. Avoid all unnecessary trips. If you must travel, use public transportation if possible. However, if you are forced to use your automobile for a trip of any distance, take these precautions:

–Make sure your car is in good operating condition, properly serviced, and equipped with chains or snow tires.

–Take another person with you if possible.

–Make sure someone knows where you are going, your approximate schedule, and your estimated time of arrival at your destination.

–Have emergency "winter storm supplies" in the car, such as a container of sand, shovel, windshield scraper, tow chain or rope, extra gasoline, and a flashlight. It also is good to have with you heavy gloves or mittens, overshoes, extra woolen socks, and winter headgear to cover your head and face.

–Travel by daylight and use major highways if you can. Keep the car radio turned on for weather information and advice.

–Drive with all possible caution. Don't try to save time by travelling faster than road and weather conditions permit.

–Don't be daring or foolhardy. Stop, turn back, or seek help if conditions threaten that may test your ability or endurance, rather than risk being stalled, lost or isolated. If you are caught in a blizzard, seek refuge immediately.

• KEEP CALM IF YOU GET IN TROUBLE. If your car breaks down during a storm, or if you become stalled or lost, don't panic. Think the problem through, decide what's the safest and best thing to do, and then do it slowly and carefully. If you are on a well-traveled road, show a trouble signal. Set your directional lights to flashing, raise the hood of your car, or hang a cloth from the radio aerial or car window. Then stay in your car and wait for help to arrive. If you run the engine to keep warm, remember to open a window enough to provide ventilation and protect you from carbon monoxide poisoning.

Wherever you are, if there is no house or other source of help in sight, do not leave your car to search for assistance, as you may become confused and get lost.

• AVOID OVEREXERTION. Every winter many unnecessary deaths occur because people—especially older persons, but younger ones as well--engage in more strenuous physical activity than their bodies can stand. Cold weather itself, *without* any physical exertion, puts an extra strain on your heart. If you add to this physical exercise, especially exercise that you are not accustomed to—such as shoveling snow, pushing an automobile, or even walking fast or far—you are risking a heart attack, a stroke, or other damage to your body. In winter weather, and especially in winter storms, be aware of this danger, and avoid overexertion.

Chapter 5
EARTHQUAKES

If your area is one of the places in the United States where earthquakes occur, keep these points in mind:

—When an earthquake happens, *keep calm*. Don't run or panic. If you take the proper precautions, the chances are you will not be hurt.

—REMAIN WHERE YOU ARE. If you are outdoors, stay outdoors; if indoors, stay indoors. In earthquakes, most injuries occur as people are entering or leaving buildings (from falling walls, electric wires, etc.).

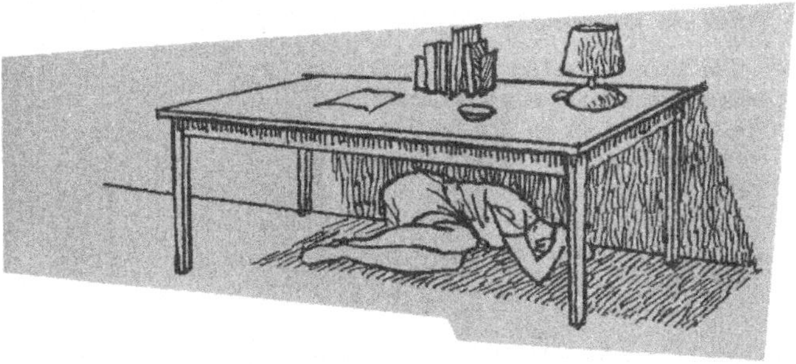

—If you are indoors, sit or stand against an inside wall (preferably in the basement), or in an inside doorway; or else take cover under a desk, table or bench (in case the wall or ceiling should fall). Stay away from windows and outside doors.

– If you are outdoors, stay away from overhead electric wires, poles or anything else that might shake loose and fall (such as the cornices of tall buildings).

– If you are *driving an automobile,* pull off the road and stop (as soon as possible, and with caution). Remain in the car until the disturbance subsides. When you drive on, watch for hazards created by the earthquake, such as fallen or falling objects, downed electric wires, and broken or undermined roadways.

After an Earthquake

For your own safety and that of others, you should follow carefully the advice given in the section, "After a Natural Disaster".

Made in the USA
Monee, IL
19 October 2026

66036551R00059

ARTETERAPIA HUMANISTA

Proceso gestáltico a través de los chakras

Colección Gestalt Dirigida por **JUAN JOSÉ DÍAZ**

LA GESTALT, desde sus influencias precursoras, tiene casi un siglo de existencia. En su devenir ha discurrido por contextos geográficos, políticos y científicos que han modulado su evolución. Entre los gestaltistas hay cierto consenso en los acontecimientos y etapas que definen su historia, en el valor de personas y textos singulares que le dieron realce, y en el reconocimiento de movimientos sociales, culturales y profesionales que diversificaron su expresión en Tendencias y Escuelas. Hoy, la Terapia Gestalt está extendida y asentada por el mundo. Y en continua adaptación y cambio.

Queremos atender las necesidades de los lectores de habla española, en la dirección de completar un abanico cada vez más amplio de textos en nuestro idioma, lo que equivale a aumentar la diversidad de pensamiento, de creación y de libertad en el arte de la terapia. Esperamos así que nuestros libros permitan acceder a textos desconocidos de autores clásicos, integrar la obra de otros gestaltistas poco conocidos, y en general recoger contribuciones a una Gestalt flexible y en continua renovación.

La Colección Gestalt de Mandala Ediciones emerge, del fondo de publicaciones existentes de este enfoque, queriendo llenar un vacío de documentos audiovisuales que nos muestren de una forma directa cómo es en la práctica la Terapia Gestalt. Nuestra intención es que este medio de comunicación, presentado en DVD, acerque aún más la Gestalt a un público general, sirva de referencia a estudiantes y dé a los profesionales nuevas herramientas de reflexión sobre su actitud y oficio.

Deseamos que nuestras publicaciones, impresas o en DVD, contribuyan a dotar a las Escuelas de Gestalt de material didáctico de calidad.

Entre nuestros propósitos consideramos con especial cariño el facilitar y apoyar la difusión, en documentos impresos y audiovisuales, del quehacer de los terapeutas gestálticos de habla española. Y, en la medida de lo posible, hacer llegar sus obras al foro gestáltico internacional.

Éste es nuestro compromiso.

Colección Gestalt

COLECCIÓN GESTALT
Director de la colección: Juan José Díaz
E-mail: gestaltmandala@gmail.com
http://www.gestaltmandala.com

Colección nº 9:
Arteterapia humanista.
Proceso gestáltico a través de los chakras

Tarragona 23 • 28045 Madrid (España)
Tel.: +34 914 678 528
E-mail: info@mandalaediciones.com
www.mandalaediciones. com

Fotografía de portada: "Manos modelando un corazón en arcilla"

Diseño gráfico: reiko

Maquetación: Torre, S. L.

Printed by

I.S.B.N.: 978- 84-8352-411-4

Depósito Legal:

COLECCIÓN Gestalt
Nº 9

ARTETERAPIA HUMANISTA

Proceso gestáltico a través de los chakras

Elvira Gutiérrez Rodríguez

MANDALA EDICIONES

AGRADECIMIENTOS

A David y a mis amigos, por su existencia.

Agradezco a mis clientes y alumnos, todo lo que juntos hemos aprendido.

Agradezco a todos los profesionales que colaboran en los cursos de Arteterapia que, entendiendo y asumiendo el programa en su conjunto como un proceso que requiere de etapas, y compartiendo el criterio de crecer y aprender con los alumnos, han aportado sus conocimientos específicos y su personalidad.

Agradezco a todos aquellos con los que me he formado, Carmen Durán, Claudio Naranjo, Yuji Yahiro, Rempo San, Wayne Liquorman, Paco Peñarrubia, y muchos más; y también a mis padres, y a aquellos a los que no he podido conocer salvo por sus escritos y que también me han dejado su huella.

Agradezco a Juanjo Díaz, su cuidadosa lectura, ideas y correcciones.

Agradezco a la vida tener una humilde casita en la Alpujarra, sin internet, ni tele, ni distracción alguna salvo unas magníficas montañas que cambian de color, donde he podido refugiarme a escribir.

Índice

Prólogo

El sistema pedagógico que recojo en este libro es el resultado de un recorrido personal que me ha ido llevando a entender al ser humano como algo muy rico en matices y en potencialidades, a la vez que complejo en su funcionamiento.

Estoy hablando de mi propio recorrido. Comencé a estudiar Bellas Artes teniendo ya un trabajo y una familia propia con un hijo pequeño, mi motivación era la necesidad de expresar sentimientos y vivencias que no tenían cabida en la palabra; lo intenté por mi cuenta, no fui capaz y pensé que si hacía una formación académica, nada menos que una carrera, podría ser capaz de expresarme con esas herramientas. Pero lo que me enseñaron, nada tenía que ver con ello. Ahora sé, que mi expresión a través de las obras de arte sí se producía, pero no sabía reapropiarme de la obra después, con lo cual algún beneficio que yo esperaba, no llegaba a producirse. En paralelo, la práctica del judo me iba haciendo recuperar un contacto perdido con mi cuerpo.

Más tarde llegaron, el yoga, la psicoterapia gestalt y años de práctica de meditación zen, y todo aquel combinado fue cuajando; empecé a ver un sentido unitario a mis prácticas, a mi vida. Recuerdo la sensación de que el puzzle empezaba a encajar.

Cuando llevaba años dando clases de pintura, veía cómo muchas de las dificultades aparentemente técnicas de mis alumnos, no lo eran, sino que simplemente dejaban traslucir bloqueos corporales o conflictos emocionales, por lo que empecé a proponerles que realizasen algunas actividades que yo creía que les podían ayudar; pero los beneficios resultaban solo parciales. Luego entendí que faltaba que todo ello estuviese integrado de alguna manera, era como una mahonesa con todos los ingredientes pero sin ligar.

Fue entonces cuando comencé a proponer cursos; al principio fueron seminarios, donde se aunaban los ingredientes gracias a la colaboración de diferentes profesionales, dando lugar a unas experiencias que iban más allá del arte, del trabajo corporal, de la psicoterapia, de la dinámica grupal y de la meditación, que eran todo eso, y otra cosa, era arteterapia.

Aquello dio lugar a que escribiese la primera tesis doctoral sobre arteterapia en España y sobre todo creó los fundamentos de la escuela, que con el tiempo ha ido evolucionando y creciendo gracias a los profesores que compartiendo las ideas esenciales, han aportado sus diferencias.

Entiendo que un sistema tan complejo como el ser humano, que funciona mediante el sistema de compensaciones, que cuando un pie pisa mal todo el resto del cuerpo se reajusta para equilibrar, no puede ser ni tratado ni comprendido si no es de una forma unitaria, y eso significa atendiendo en paralelo al cuerpo físico, mental, emocional, y espiritual. Eso es Arteterapia Humanista.

Ahora, después de 15 años de práctica en sesiones individuales y grupales, me planteo el reto de presentarlo ordenado y por escrito, lo cual tiene un punto paradójico, ya que la metodología de este sistema pedagógico está basada en la observación de lo que ocurre en el aquí y ahora, y en las relaciones que se van creando con los intercambios que surgen. Aún así, considero que vale la pena el intento de transmitir la idea esencial de un trabajo de desarrollo personal que contempla al ser humano como una unidad; que a su vez está formando parte de unidades mayores, familia, grupo, humanidad, cosmos; y que contempla la sanación como la eliminación de lo que sobra, de lo que en su momento hizo falta para la supervivencia pero que ya no se requiere y nos dificulta la existencia; como la recuperación, también, de cualidades intrínsecas como la curiosidad y la afectividad, presentes cuando niños y muchas veces olvidadas cuando mayores; para así poder acceder al Ser interior, y desde ahí darse cuenta de que ni los juicios, ni los miedos, ni la culpa, ni la soledad tienen sentido, cuando se entiende que el Ser es uno.

Entiendo la espiritualidad, por tanto, como algo inherente al ser humano, y al objetivo del desarrollo personal como la conexión con lo más auténtico de uno mismo, para desde ahí tener un contacto genuino con el otro, de mi Ser con el otro Ser para abrir la posibilidad al reconocimiento de la unidad y por tanto de lo transcendente.

Introducción 00

Mi deseo más ferviente es que quien lea este libro entienda el Arteterapia Humanista como un PROCESO de crecimiento. A pesar de que se habla de métodos y técnicas, creo sinceramente que no existen ejercicios mágicos, que es la escucha y acompañamiento de la persona o del grupo, con las interacciones que surgen del contacto real, lo que posibilita el desarrollo, tanto del cliente como del arteterapeuta; y para que este contacto real se produzca, es necesaria la presencia plena, objetivo final de este proceso.

Es necesario entender la importancia del acompañamiento a lo largo de un viaje que necesariamente pasa por fases de ilusión, desilusión, desesperación, reencuentro del propio ser, aceptación, etc.., y que el arteterapeuta conoce, porque las ha transitado él mismo, y es por ello que puede orientar y dar confianza durante el recorrido.

La realización de algunos de estos ejercicios de forma aislada, puede parecer eficaz en cuanto que son susceptibles de desvelar zonas semiocultas de la personalidad, diríamos que pueden abrir una rendija en una puerta anteriormente cerrada, pero es una rendija, tiene más que ver con la satisfacción de la curiosidad que con un desvelamiento, ya que para que éste se produzca es necesario que afecte profundamente a diferentes partes de nuestro ser, y así nos transforme. Este es el objetivo que nos proponemos.

EJES DE MI PEDAGOGÍA

Para que el cambio profundo se produzca, mi pedagogía se basa en tres ejes.

— **Eje vertical**: consiste en un recorrido a lo largo del cuerpo, siguiendo el orden de los chakras, con la aplicación de

- técnicas corporales,
- expresivas y
- de análisis,

con el objetivo de actuar en paralelo en

- el cuerpo físico,
- emocional,
- mental y
- espiritual de la persona.

— **Eje horizontal**: tiene que ver con el dar y recibir, con lo relacional. Y afecta por ello a:

- el trabajo individual, autorresponsabilidad por el propio proceso
- la inter-relación grupal, por medio de las interacciones, proyecciones y resonancias que se dan
- el entorno familiar y social, donde se ven en el día a día las acciones, reacciones y cambios
- la relación transferencial, con tutores y terapeutas, espejo de la evolución del proceso.

— **Eje en profundidad**: El trabajo va ejerciendo su acción capa por capa de la estructura egoica:

- aumentar la auto-confianza,
- atravesar las corazas, los bloqueos, las resistencias
- descubrir las sucesivas máscaras del ego
- poder entrar en contacto con el Ser interior, y desde ahí
- establecer un contacto genuino con los otros

Estos tres ejes se apoyan entre sí, formando una malla que va sosteniendo a la persona y su proceso.

Al mismo tiempo la metodología sigue un doble camino:

— El Proceso Personal: con los tres ejes explicados
— La Formación en métodos y técnicas, así como la ejemplificación en vivo y la exposición de conceptos teóricos

Las propias características de este enfoque pedagógico basado en la observación de lo que ocurre en el presente y en las relaciones, hace difícil su transmisión por escrito, por ello lo que más se evidencia en el texto es el trabajo con el eje vertical, pero entiéndase que todo el tiempo los otros dos ejes están también presentes.

CONSIDERACIONES RESPECTO A LA LECTURA DE ESTE LIBRO

— Masculino-femenino: Un problema con soluciones poco satisfactorias: se me hace pesada la lectura de textos que insisten en cada palabra en incluir el masculino/femenino. Tampoco me gusta caer en el lenguaje sexista. El uso de la "e" como alternativa, aunque creativa, me distrae de la comprensión. He optado por una decisión salomónica, mitad y mitad; unos capítulos están en masculino y otros en femenino. Me parece equitativo y menos distorsionante. Espero que unos y otras entiendan mi buena voluntad.

— El uso del presente y de la primera persona: En la Universidad y en los textos que quieren ser serios, es imperativo el uso de la tercera persona, es más impersonal y da por tanto una apariencia mas científica, me encanta usarla, pero "se insiste", no, yo también lo hago, insistimos tanto, en Gestalt como en Arteterapia Humanista, en responsabilizarnos, en hablar en primera persona, que me parecería una incongruencia no hacerlo. Así que durante todo el texto hay un baile entre la primera y la tercera persona que tiene su sentido, en separar lo general de lo particular, y que espero que le de un cierto movimiento.

— Todos los ejemplos que se emplean, tanto con imágenes como con comentarios, para ayudar a la comprensión de conceptos, se realizan con permiso de los implicados y respetando su intimidad; los nombres utilizados han sido cambiados.

En el primer capítulo hago una exposición de lo que es el Arteterapia Humanista, en qué se apoya el arteterapia, sus antecedentes, diferencias entre las diferentes tendencias en arteterapia, y entrando ya en lo que es la corriente Humanista, sus características; también explico el por qué del trabajo con los chakras, y qué lugar ocupa la espiritualidad en este enfoque; terminando con el papel de la creación artística y la importancia del *setting*.

En el segundo capítulo hablo de la terapia Gestalt, de su ubicación dentro del mundo de las psicoterapias, y de la importante relación que ha mantenido con el arte y los artistas desde sus inicios. También de cómo en el presente, de la mano de algunos artistas y terapeutas, ambas disciplinas se ayudan mutuamente.

Estos dos primeros capítulos son la parte más teórica del libro, algunas personas quizás prefieran dejarlo para el final, pero para quienes sea su primera aproximación al tema puede ayudarles a ubicar de lo que se está hablando.

En el tercer capítulo explico brevemente el recorrido a través de los chakras tratando de transmitir el sentido de unidad que este recorrido tiene.

Los siguientes siete capítulos, del 4 al 10, contienen la explicación chakra por chakra de la referencia corporal, contenidos emocionales que se tratan en cada uno de ellos, ejercicios, acciones artísticas que se plantean, su justificación, y breve explicación metodológica, así como algunos ejemplos vivos de las respuestas a las propuestas hechas. Terminando cada uno de ellos con un microrrelato alusivo al tema, para abordarlo también desde un lenguaje metafórico o analógico.

Termina con el capítulo de conclusiones en que se hace hincapié en que ser arteterapeuta no consiste en conocer una serie de técnicas y métodos sino que es necesario haber realizado un proceso personal de auto-indagación y desarrollo que permita afrontar el acompañamiento de otros con empatía, con un talante humanista y con seguridad porque se ha recorrido el camino, y se conocen los baches y las cuestas.

Arteterapia Humanista 01

❖ QUÉ ES ARTETERAPIA

Consiste en la utilización de medios artísticos en un contexto terapéutico, o de autoconocimiento y desarrollo personal.

Arteterapia no es la unión de 2 herramientas diferentes: Arte y Terapia; tampoco es la alternancia entre los dos procesos artístico y terapéutico; ni es que el proceso artístico actúe como detonante y el otro como terapéutico; sino que ambos se apoyan entre sí y juntos constituyen el encuadre arteterapéutico. El medio artístico y el medio terapéutico pierden su hegemonía para configurar el espacio arteterapéutico.

El Arteterapia tiene lugar dentro del *Campo* que crean Cliente/Arteterapeuta/Experiencia artística. Esta "triangulación" propicia que el creador o cliente, sienta la curiosidad de indagar sobre lo que su experiencia estética ha creado.

Existen tantas corrientes como corrientes psicológicas hay: Conductista, Psicoanalista, Sistémica, Humanista...

La metodología varía en función de la corriente psicológica a la que se adscriba el arteterapeuta y del sujeto o grupo en quien se aplique.

La terapia artística se fundamenta en que todas las personas, tengan o no tengan cualidades artísticas, poseen la capacidad para proyectar bajo formas visuales su mundo interno, tanto conflictos como intuiciones o inspiraciones.

Los trabajos pueden ser espontáneos o bien haber sido provocados por una propuesta de trabajo del terapeuta artístico, lo que da lugar a diferentes estilos terapéuticos, más o menos directivos, que se adecúan a diversos tipos de grupos de clientes.

Un factor muy importante del Arteterapia lo constituye el hecho de que no tiene que ser necesariamente verbal, lo que extiende su campo de aplicación, tanto a niños, como a adultos a los que les resulte difícil expresar sus estados emocionales, y pensamientos, a través del lenguaje verbal.

El lenguaje verbal está enfocado al razonamiento discursivo, mientras el lenguaje no verbal, tanto corporal, plástico, del sonido... puede abarcar otros espacios mas amplios; y además está el hecho de que en ocasiones, este tipo de lenguaje verbal puede ser empleado fácilmente como defensa; es muy fácil escudarse en el lenguaje verbal cuando éste se domina, para racionalizar el proceso y salirse de la experiencia.

Otro beneficio añadido del empleo de medios artísticos en la terapia, lo constituye el hecho de que provee de una memoria de la sesión o sesiones en que se ha producido, lo que posibilita la vuelta a ese punto, bien para una relectura posterior, o bien para continuar trabajando desde el punto en que se dejó.

ANTECEDENTES DEL ARTETERAPIA

El Arteterapia surge de la unión de dos escuelas de pensamiento, el Psicoanálisis y la Educación Artística. Comienza a practicarse hacia los años 40, basándose en el resultado terapéutico de la actividad creadora y, en gran parte, alentado por el interés que la cultura de la época tiene por el Surrealismo y la valoración que esta corriente cultural hacía de las imágenes inconscientes, ya fuesen éstas de origen onírico, productos de un buscado automatismo, o sugeridas por el opio y la absenta; este clima, conllevó un interés por las obras de los artistas psicóticos, haciendo que éstas comenzasen a ser valoradas por el personal médico; consecuentemente, llevó emparejado que se generalizasen los talleres de arte en los hospitales y clínicas psiquiátricas.

Este interés originó que surgiese una nueva profesión. Al comienzo fueron los artistas o los terapeutas ocupacionales, los encargados de estos talleres, más tarde se interesaron algunos psiquiatras por esta labor, y poco a poco comenzó a estructurarse este nuevo campo como una especialidad de las enseñanzas de Bellas Artes en unas ocasiones y en otras, como una nueva disciplina a caballo entre las Bellas Artes y la Psicología.

Los antecedentes de las dos escuelas de pensamiento que influyeron en el Arteterapia desde sus comienzos los podemos situar en:

- Naumberg que influenciada por el psicoanálisis utiliza el arte dentro de la relación psicoterapéutica,

– Kramer que enfatizó las propiedades curativas de la actividad artística en sí misma, sin necesidad de interpretación.

Hasta llegar a la decisiva aportación de Winnicott, D. (1971) para el Arteterapia (con sus estudios sobre el "objeto transicional", a medio camino entre el yo y el ello, lugar de ubicación de la práctica artística y del objeto artístico), hay un corto pero intenso recorrido, y digo intenso, porque como corresponde a nuestra época, es un recorrido convergente y multidisciplinar de diversas ramas de la ciencia que confluyen en mirar el hecho artístico y la creatividad, como una posible vía de curación del ser humano.

Entre las primeras investigaciones clínicas sobre el tema hay que citar las que Anastasi y Foley (1944) realizaron sobre el diagnóstico clínico de adultos psicóticos. Respecto a las investigaciones psicológicas, los antecedentes se pueden situar en Machover (1949) sobre el dibujo de la figura humana, y Lowenfeld (1947) sobre el dibujo infantil.

Entre las investigaciones realizadas desde una óptica relacionada con el arte, cabe mencionar los estudios de Gombrich(1963) sobre el psicoanálisis en la historia del arte; los que realizó Cirlot (1949) sobre "Infantilismo" en el arte; los trabajos de Arnheim (1969) sobre el "Pensamiento Visual" en los que nos dice que "pensar con los sentidos" puede ser parte de un proceso terapéutico, insistiendo en que, los fenómenos físicos y psicológicos son isomórficos -sus formas son estructuradas de modo similar-, idea que recoge y amplia el Arteterapia considerando que las imágenes que el cliente crea son isomórficas con sus patrones de comportamiento.

Así como, hay que mencionar también, el interés, coleccionismo y escritos de varios renombrados artistas por lo que fue llamado "el arte bruto" de los psicóticos, entre estos artistas están Max Ernst, Dubuffet y Breton, para todos ellos la expresión artística de los enfermos mentales venía a ser el modelo de un arte creador sin trabas, sin represiones.

Entre los médicos que comenzaron a reunir y conservar las obras de sus pacientes destaca la famosa colección Prinzhorn (1880-1920), hoy en la Clínica Psiquiátrica Universitaria de Heidelberg.

TIPOS DE ARTETERAPIA

Puede hacerse una descripción de los diferentes tipos de Arteterapia atendiendo a diversos factores:

— Vinculación teórica: conductismo, psicoanálisis, sistémica, humanismo...
— Grupo de clientes a quien se dirija: individuos o grupos; grupos de apoyo, de reinserción, hospitalarios...
— Función que se asigne a la obra: producto final, proceso de ejecución, objeto intermediario...
— Estilo del arteterapeuta: directivo, no directivo.

Según la corriente a la que esté adscrita:

— Se atiende más o menos, al producto final.
— Se atiende o no, al proceso de ejecución.
— Se trabaja sobre las relaciones transferenciales que tienen lugar.
— Se realizan interpretaciones y diagnóstico en base a la obra, o se ayuda al autor a su auto-lectura.
— Se pone el énfasis en la adquisición de técnicas y comportamientos, o pone el énfasis en la auto-indagación.
— Se hace acompañamiento y apoyo solamente, o se hace también confrontación.

❖ QUÉ ES ARTETERAPIA HUMANISTA

Puede considerarse a Jung (1875-1961) precursor del Arteterapia Humanista por diversas razones:

— Establece una analogía entre la obra de creación artística y el proceso de individuación.
— Preconiza la imaginación activa como material salido del inconsciente, que es factible expresar plásticamente, por escrito o mediante cualquier otra forma, y que va a permitir durante el tratamiento un ir y venir entre lo no verbal y lo verbal, el inconsciente y la conciencia.
— Entiende la imaginación creadora como personal y suprapersonal, dándole así un carácter transcendente.
— Durante años llevó un diario personal de mandalas, que entendía que ayudaban a dar unidad a la psiquis.

El Arteterapia Humanista entiende con Maslow (1908-1970), uno de los creadores de la Psicología Humanista, que la salud no es sólo la ausencia de enfermedad y que un ser humano sano es aquel ser integrado, que

desarrolla al máximo todo su potencial de cuerpo, mente, emociones y espíritu.
Se apoya en el enfoque y las técnicas de otras psicoterapias humanistas, tanto de la Bioenergética con su énfasis en lo corporal-emocional, como en la terapia Gestalt con su enfoque en el Darse Cuenta en el aquí y ahora, en el concepto de Responsabilidad, compartiendo todas la convicción de que el ser humano tiende a su autorrealización en el aquí y ahora, con los otros y con el mundo.

CORRIENTES que han influido en el ARTETERAPIA y otras Psicoterapias Humanistas:

- El psicoanálisis de Freud, retomando y reformulando su teoría de los mecanismos de defensa y el interés por el inconsciente.
- La filosofía existencial, de la que rescata la confianza en las potencialidades inherentes al individuo, el respeto a la persona y la responsabilidad.
- La fenomenología, de la que toma su apego por lo obvio, por la experiencia inmediata y por la toma de conciencia o *insight.*
- La psicología de la Gestalt, con su teoría de la percepción (figura-fondo).
- El budismo Zen, con su búsqueda de la presencia total en el aquí y ahora.
- El psicodrama, de J. L. Moreno, del que adopta la idea de dramatizar las experiencias y los sueños.
- La teoría de la coraza muscular de W. Reich.
- La teoría de la Indiferencia Creativa, de Salomo Friedlaender, de la que extrae su teoría de las polaridades.
- La teoría de Campo por cuanto entiende lo psíquico como una zona de interacción entre organismo y ambiente, o una zona de encuentro entre sujeto y objeto como Lewin y Goodman, defienden.
- El concepto Jungiano del "sí mismo" para designar la totalidad de consciente e inconsciente del ser humano.

Se citan estas referencias, por quienes puedan estar interesados en las fuentes de las que han bebido las psicoterapias humanistas y en concreto el Arteterapia, pero es importante decir que es la integración de todas ellas (como se explica en la Introducción de este libro) en una red o malla que las contiene y contempla con una visión holográfica, lo que les da el sentido y su función en Arteterapia Humanista.

ALGUNAS CARACTERÍSTICAS QUE DIFERENCIAN EL ARTETERAPIA HUMANISTA DE OTROS ENFOQUES DE ARTETERAPIA

- Centra la atención en el darse cuenta del aquí y ahora
- La referencia corporal, la atención a las sensaciones corporales como desencadenante
- Mantiene la idea de proceso en marcha del que van emergiendo formas, situaciones cambiantes. No cristalización.
- Se respeta y valoriza la responsabilidad del cliente en dicho proceso, es quien elige y decide en todo momento.
- Se focaliza la atención sobre el proceso más que sobre el objeto creado.
- No se interpreta sino que se atiende a lo que está ocurriendo en el presente
- No se admiten los juicios de valor sobre obras o actitudes de otros.
- Se entiende que existe un elemento central de relación para el desarrollo del yo y es el Contacto, entendiéndolo como "presencia plena" en el encuentro con el terapeuta y/o el resto del grupo.
- Se centra en lo que fluye o se evita en lugar de en la transferencia o la contratransferencia.

LOS CHAKRAS EN ARTETERAPIA HUMANISTA

Los chakras son algo no tangible, no demostrable por la ciencia en el punto de conocimientos actuales, pero durante siglos, en Oriente, han sido la base del estudio del ser humano, de su fisiología, psicología y espiritualidad, y al conceptualizarlo por separado, soy consciente de que dejo escapar su esencia, que es la comprensión de la unidad y la interrelación entre todo el sistema.

Es precisamente por la visión holística que supone la referencia a los chakras, por lo que los tomo como mapa para marcar con ellos un recorrido que, necesariamente va a contemplar todos los factores que componen el individuo; apoyándose en primer lugar, en una referencia física y corporal, un espacio en el cuerpo.

Entendemos, desde el Arteterapia Humanista, que en el principio de toda experiencia está el cuerpo, lo que se experimenta corporalmente, lo que nos permite percibir, movernos, sufrir, desear y gozar. Decir pues, que la experiencia proviene básicamente del cuerpo, equivale a decir que el pensamiento, la emoción o el sentimiento, la creación artística, el comportamiento, la cognición, el inconsciente incluso, tienen su punto de partida en las sensaciones corporales. Sobre este concepto Merleau-Ponty, M. (1985) defiende que el cuerpo no es lo que se opondría al espíritu, sino que, el cuerpo es la unidad del ser. Esta facultad que nos ofrecen los chakras de poder tratar temas que afectan a toda la psiquis del individuo, manteniendo un anclaje en el cuerpo, dota a la experiencia de una solidez y tangibilidad muy deseada, por cuanto evita la dispersión mental y ayuda a su integración.

Existen diversas técnicas para trabajar con los chakras en diferentes tradiciones, pero quizás lo más importante que aporta este método, es la combinación de técnicas cada una dirigida expresamente al cuerpo mental, cuerpo emocional, cuerpo físico, y todas ellas apuntando más allá, hacia una dirección común, integradora que no olvida la parte espiritual del ser.

Añadir solamente que los mapas, por su propia naturaleza, son simplificadores, y que cualquier tema que tratemos aunque abordado desde un determinado chakra, tiene ramificaciones e interconexiones con otros temas y otros chakras; pero didácticamente, es útil compartimentar en una primera aproximación, hasta comprender el funcionamiento de aquello que se trata; posteriormente, y en nuestro enfoque pedagógico también lo hacemos, llega el momento de encontrar los vínculos y relaciones entre mecanismos y chakras diferentes.

LA ESPIRITUALIDAD EN ARTETERAPIA HUMANISTA

No existen chakras terrenales y espirituales, no hay separación; el ser humano está compuesto por huesos, músculos, piel, emociones, sensibilidad, sentido de transcendencia, etc.; tengamos la piel más áspera o más fina, los músculos más o menos desarrollados, nuestra composición es la misma, y en todos está el sentido de inmanencia, no en un cajón aparte, sino que forma parte de todo nuestro ser, desde el cuerpo y la sexualidad, al silencio o la angustia del vacío. Es con esta premisa desde donde debe entenderse el recorrido a través de los chakras con el Arteterapia.

Jung en su libro *Psicología y Religión* (1949), explica con casos tratados por él, que las experiencias religiosas existen, y a quien las ha tenido no se le puede decir que son una fantasía; son experiencias muy vívidas que no dejan lugar a dudas sobre su realidad, otra cosa es que sean irracionales. Ahí es donde Jung justifica su método de intervención como fenomenológico, ya que trata de sucesos, acontecimientos y experiencias que, tanto se califiquen de racionales o irracionales, constituyen la experiencia que el cliente ha vivido y sobre ello se trabaja.

"La experiencia religiosa es absoluta", continúa diciendo Jung, señalando que no se trata de una cuestión de fe, sino de *experiencia*. No cabe discutir acerca de ella. Una persona puede decir tan sólo que nunca tuvo una experiencia de esa índole, a lo cual replicará el opositor: Lo lamento mucho, pero yo sí.. Y ello pondrá término a toda discusión. "(...) *Es indiferente lo que piensa el mundo en punto a la experiencia religiosa: quien la tiene, posee, como inestimable tesoro, algo que se convirtió para él en fuente de vida, sentido y belleza, otorgando nuevo brillo al mundo y a la humanidad..*"

Por su parte, Butelman, en el prólogo del mismo libro a la edición castellana argumenta la diferencia de visión de Jung con respecto a Freud, diciendo: *"Condicionado históricamente Freud ve como su época le obliga a ver. En otras palabras, no pudiendo liberarse del materialismo científico de fines del s. XIX, concibe lo inconsciente de modo exclusivamente racionalista e intenta esclarecer toda creación espiritual compleja con arreglo a su imagen mecanicista del universo. De ahí su teoría de la sublimación y el considerar toda cultura humana como mera derivación del instinto sexual. De ahí también su valoración negativa de los factores irracionales.*

Lo irracional, empero, existe; es un hecho psicológico, por lo tanto, como tal hemos de considerarlo, cuidando de no violentar su idiosincrasia."

Suscribo totalmente esta actitud respecto al tratamiento de las experiencias espirituales; no caer en un escepticismo negador, que puede llevar a calificar de psicótico a todo aquel que haya tenido experiencias que al profesional, hasta el momento le sean desconocidas; ni caer tampoco en el extremo de, si son de índole espiritual, ya no pueden ser contempladas por la psicoterapia porque la sobrepasan. Cualquier experiencia, afecta a todo el campo del individuo y como tal debe de ser respetada y es susceptible de ser explorada.

Otro aspecto importante de la espiritualidad en Arteterapia lo da el hecho de considerar el encuentro cliente-arteterapeuta como una oportunidad para que lo no previsto por ninguna de las partes suceda; el abrirse a la experiencia y fluir con ella confiando en lo que está más allá de nuestros egos.

LA CREACIÓN ARTÍSTICA EN ARTETERAPIA HUMANISTA

Hablamos de creación artística, de expresión artística o de acción artística, y no de obra de arte, porque lo que nos importa en primer lugar, es la sucesión de gestos y acciones que se van dando, y que constituyen el material de trabajo; el resultado final ofrece también interés, como sugeridor de respuestas al cliente, pero no en atención a valoraciones técnicas en la factura de ejecución.

No es por tanto necesario, conocimiento previo sobre las técnicas artísticas con que se trabaja, más allá de lo más básico que suele formar parte del calentamiento previo a la actividad y que se integra ya en la indagación personal.

Esta transformación de todas las artes, en vías de conocimiento, unido a la valoración social de expresiones artísticas como el *art-brut, arte povera, naïf,* o los *happenings* e instalaciones, plantea un interrogante epistemológico, sobre el sentido de las reglas, a veces rígidas y sujetas a modas, de ejecución de cada una de las artes. Aunque éste es otro tema, lo menciono, dada la inhibición para la expresión que supone para muchas personas saberse sin esos conocimientos, que repito no son necesarios.

Al emplear como vehículo de conocimiento las expresiones artísticas lo que estamos proponiendo no es solamente un nuevo lenguaje, es unir la realidad interior y exterior dejando espacio para lo imaginario y lo simbólico, y todo ello de una manera no planeada previamente y por tanto limitada, sino permitiendo que la acción esté viva, que sorprenda a quien la ejecuta en su evolución, y para que esta libertad de juego exista, es indispensable que no esté condicionada por el resultado final.

Utilizando de esta manera los lenguajes artísticos (plástica, teatro, danza, música...), estos pueden convertirse en un medio para:

- La expresión y comunicación desde otros registros diferentes del habitual.
- Exteriorizar situaciones conflictivas difíciles de verbalizar.

- Experimentar la creatividad encontrando soluciones nuevas para la resolución de situaciones.
- Entrar con facilidad en espacios mentales nuevos, quizás no lineales, o no racionales.
- Abrirse a información procedente del cuerpo o del subconsciente.
- Acceder a lo espiritual que hay en cada uno de nosotros.

EL SETTING

En Arteterapia Humanista existen dos marcos:

- Un **marco externo**, como en cualquier otra relación terapéutica, dado por el encuadre, o contrato entre el participante y el arteterapeuta, incluye no solamente los acuerdos respecto a horario, pagos, tiempos..., sino una primera aproximación de lo que la persona desea trabajar. En este marco, también hay que contemplar la posibilidad de que el cliente acuda en paralelo a otra terapia o tratamiento lo que entraría a formar parte del marco de relación.
- Un **marco interno** que lo da la expresión artistica con la que se trabaja ya que cada una de ellas tiene sus propias reglas y límites. Así mismo, dependiendo de la producción artística, se acordará su conservación o destrucción.

Pero el *setting* en Arteterapia es mas que el marco, constituye un elemento activo:

- **El papel del arteterapeuta**, por ej., puede variar en función de la necesidad del momento, y pasar de ser testigo, a participar activamente; el espacio físico que ocupa en la sala, a diferencia de en otras terapias, puede cambiar incluso varias veces durante la sesión; o el contacto físico, en otras terapias vetado, puede entrar a formar parte de la interacción, siendo todos estos factores importantes del proceso terapéutico que afectan al *setting*.
- **El espacio de la terapia** es también determinante; promover el contacto o el movimiento libre, en un aula con mesas y sillas es, cuanto menos, paradójico; proponer la expresión en arcilla en un despacho con diván, no parece fácil; por otro lado la posibilidad de convertir la sala o despacho temporalmente en una sala de exposición de la obra o en escenario de actuación, hace del espacio de la terapia un elemento facilitador de la misma.

- **La obra,** como objeto transicional entre la realidad exterior e interior (Winnicott, 1971) forma parte del *setting,* pero es, al mismo tiempo, un elemento de la terapia y una proyección, casi diría prolongación, del cliente.

Concluyo, en Arteterapia Humanista, la sesión incluye todo aquello que sucede desde que se abre hasta que se cierra la puerta de la sala o consulta, desde el buenas tardes hasta el adiós, incluyendo la relación con el mobiliario, materiales...

Gestalt 02

❖ QUÉ ES GESTALT

Gestalt es un término alemán, sin traducción directa al castellano que aproximadamente significa "forma", "totalidad", "configuración".

La forma o configuración de cualquier cosa está compuesta de una "figura" y un "fondo". Por ejemplo, en este momento para quien lee este texto, las letras constituyen la figura y los espacios en blanco forman el fondo; aunque esta situación puede invertirse y lo que es figura puede pasar a convertirse en fondo.

Esta idea fue la base de numerosas investigaciones sobre la psicología de la percepción que dieron lugar a la Psicología Gestalt o Psicología de la Forma, como también se la conoce, una escuela teórica que hizo un importante aporte al estudio de los fenómenos perceptivos.

LA PSICOLOGÍA GESTALT

Surgió en Alemania y tuvo su apogeo en EE.UU. en los años 30. En su momento, implicó una revolución de las ciencias psicológicas al considerar los fenómenos perceptivos como unidades organizadas, en lugar de verlos como una acumulación de distintos datos sensoriales.

El núcleo de la Psicología Gestalt gira en torno a esta idea, no percibimos conjuntos de elementos sino unidades de sentido estructuradas. Por ej.: Si escucho parte de una frase tenderé a acabarla, si veo una escena o parte de ella asociaré unos significados y si una figura está incompleta el perceptor la completará en su mente.

En la experiencia de percepción hay una tendencia a que los miembros constituyan grupos, a que las figuras incompletas se completen, definan y precisen, y a que el campo total sea organizado en figura y fondo. Esta figura y su fondo forman lo que se llama una *gestalt* (configuración). Lo que percibimos son pues *gestalten* (figura + fondo), totalidades que se complementan, siendo el todo que forman más que la suma de sus partes.

El modelo propuesto por la Psicología Gestalt fue, en parte, y sólo en parte, el origen y la inspiración para dar nombre a una nueva forma de hacer terapia que estaba abriéndose paso: la terapia Gestalt.

LA TERAPIA GESTALT

Desarrollada en EE.UU., después de la Segunda Guerra Mundial por Fritz y Laura Perls, la terapia Gestalt hereda de la Psicología de la Forma el concepto de que la naturaleza humana se organiza en totalidades y es vivenciada por el individuo en esos términos; esta nueva mirada supuso una revolución en la forma de tratar a los pacientes que pasaron, de ser vistos de forma aislada en su enfermedad, a ser vistos dentro de un contexto familiar y social en el que el comportamiento, patológico o no, es un elemento de la relación.

El síntoma, por tanto, deja de ser enfocado de forma unilateral para pasar a contemplarse como una parte del modo de contacto o de evitación de contacto del individuo con su medio.

Con este encuadre, la terapia Gestalt pone el foco de interés en la conciencia, "el darse cuenta" de lo que está ocurriendo en el momento presente, en el "aquí y ahora", a nivel corporal, mental y emocional en el "yo-tu", es decir, en la persona y también en lo relacional.

Conseguir este contacto pleno con uno mismo y con el medio, es lo que la terapia Gestalt entiende que puede permitir al individuo tomar en todo momento las acciones que le lleven a la satisfacción de sus necesidades, con la consecuente eliminación de la angustia.

Por ello, el proceso terapéutico se basa en la toma de conciencia, por parte de la persona, de cuales son las situaciones en que pierde el contacto con sus necesidades o en qué distorsiona la percepción de la experiencia, y de qué modo lo hace, con el fin de que aumentando su conciencia, llegue a identificar sus necesidades auténticas con claridad y a tener más posibilidades de satisfacerlas; de esta manera, podrá evitar quedarse "enganchado" en situaciones sin resolver y estará libre para, concluido un asunto, poder llevar su atención a la siguiente cuestión que reclame su interés. Esto es lo que se denomina "cerrar gestalts".

Los pilares en que se apoya la terapia Gestalt y que se tratan en los capítulos 4 y 5 de este libro, son: la conciencia, el aquí y ahora, y la responsabilidad.

— LA CONCIENCIA (*awareness*), el darse cuenta de las sensaciones corporales; de la actividad mental, ya sea ésta motivada por fantasías, anticipaciones, recuerdos, racionalizaciones, etc.; y de las emociones que van apareciendo.

Esta conciencia puesta en los movimientos internos, va unida a una escucha atenta al entorno, es decir, a los estímulos que llegan del exterior, determinados por las circunstancias e individuos que forman parte de la vida de esa persona.

— EL AQUÍ y AHORA es un empeño que la terapia Gestalt comparte con la meditación Zen y otras disciplinas orientales; el objetivo es mantener el contacto con la situación que se está viviendo, y evitar que las fantasías de futuro o los recuerdos del pasado, distraigan de la experiencia.

Por ello desde la terapia Gestalt se insiste en la comunicación en tiempo presente. Esto no quiere decir que pasado y futuro sean algo carente de interés y de lo que no conviene hablar sino que, tanto los recuerdos como las ideas de futuro, se producen en el presente y se trabajan y se actualizan trayéndolos al ahora, atendiendo a los sentimientos que provocan en el presente al traerlos a la conciencia y a cómo repercuten en la experiencia actual.

— LA RESPONSABILIDAD es un término que suele emplearse habitualmente en relación con el compromiso con la tarea o con otros; pero se olvida con frecuencia que la principal responsabilidad recae sobre uno mismo, los propios actos y descuidos.

Se nos educa en depositar la responsabilidad por el propio cuidado y el de nuestros intereses (salud, economía, aprendizaje...) en otros, con el argumento de que son los expertos, y nos habituamos a responsabilizar a las circunstancias de gran parte de lo que nos sucede, sin darnos cuenta de que al delegar la responsabilidad cedemos también el poder sobre nuestra propia vida, dejamos de ser protagonistas para convertirnos en comparsas, con el sentimiento, en ocasiones, de ser víctimas pasivas.

La terapia Gestalt incide en que, queramos o no, somos responsables de nuestra vida y que asumirlo, nos abre la posibilidad de decidir y actuar sobre ella; nos convierte en el actor principal.

❖ UBICACIÓN DE LA TERAPIA GESTALT DENTRO DE LAS PSICOTERAPIAS

De la terapia Gestalt se ha dicho que es heredera del Psicoanálisis, que es Corporal y Existencial, que es una terapia Humanista y que es Transpersonal; todas estas definiciones no son excluyentes entre sí, sino abarcadoras, ya que unas incluyen a otras.

LA TERAPIA GESTALT Y EL PSICOANÁLISIS

La terapia Gestalt es claramente heredera del Psicoanálisis, tanto por lo que debe a los aportes freudianos, fundamentalmente el reconocimiento y la importancia del *inconsciente*, como por su controversia a la hora de abordarlos.

Tanto en lo que se refiere al *inconsciente* como en la forma de tratar otros conceptos como el de *transferencia* o el de *resistencia*, la terapia Gestalt no los niega, pero difiere en cuanto a su tratamiento con respecto al Psicoanálisis.

Hay que tener en cuenta que estas divergencias se produjeron a partir de un conocimiento e inmersión en el Psicoanálisis, ya que Perls se psicoanalizó durante años con Karen Horney, Reich y otros; se formó como psicoanalista, y ejerció más de 20 años como tal.

Si en el apartado anterior hablábamos de los aportes de la Psicología de la Gestalt a la terapia Gestalt, debemos incluir también los aportes del Psicoanálisis, aunque sea en su vertiente de discípulos disidentes.

LA TERAPIA GESTALT Y LAS TERAPIAS PSICOCORPORALES Y EXISTENCIALES

A veces se clasifica la terapia Gestalt dentro de las terapias psicocorporales, dado que al entender a la persona como un todo, incluye las experiencias corporales de ésta, tanto en lo observable, y susceptible por tanto de ser indicado, como en el trabajo del "darse cuenta " que el cliente lleva a cabo durante su proceso.

Con frecuencia se busca volver a captar la experiencia desde su origen, la sensación corporal, para que la experiencia se haga viva y por tanto modificable.

No quiere decir esto que pretenda ser una terapia corporal, ya que esa atención a lo corporal no está separada del resto, sino que es un elemento más en la atención a la totalidad del individuo.

Respecto a la consideración de Psicoterapia Existencial, no puede ponerse en duda que la filosofía de comienzos de siglo XX en Europa, el existencialismo, forma parte de la concepción de base gestáltica, en cuanto a anteponer la descripción de los fenómenos o hechos vividos, a las elaboraciones mentales sobre ellos.

El existencialismo encuentra el sentido y el valor del ser humano, no en conceptos abstractos sino en su cotidianeidad, en su vida y sus obras.

Esta atención a lo obvio de la existencia y por tanto de la relación con el otro, está en el fundamento de la sustitución del enfoque basado en el vínculo transferencial (freudiano) por el encuentro dialogal entre terapeuta y cliente (yo-tú). Podemos decir por tanto que el existencialismo está en la médula de la terapia Gestalt.

LA TERAPIA GESTALT Y LAS PSICOTERAPIAS HUMANISTAS

Carl Jung, desarrolló el concepto de "proceso de individuación" el cual transcendía lo personal; Roberto Assagioli aseguraba que el desarrollo de la salud adulta incluía una psicosíntesis de lo personal y lo espiritual; y Abraham Maslow, defendía que la salud no era solamente la ausencia de enfermedad sino que se produce cuando el ser humano desarrolla todo su potencial lo cual incluye la espiritualidad, la creatividad...

Habitualmente se encuadra la terapia Gestalt en el movimiento de las Psicoterapias Humanistas, ya que comparte con dicho movimiento las premisas clave:

- La visión integradora del individuo, al que entiende como unión de cuerpo, mente, emociones y espíritu.
- La confianza en la sabiduría intrínseca del organismo, es decir, de la persona en su totalidad, para autorregularse y sobrevivir en un medio y circunstancias cambiantes.
- La convicción de que el ser humano tiende a su autorrealización en el aquí y ahora, con los otros y con el mundo.

De todo lo cual se desprende una visión del ser humano que requiere de un acompañamiento que, como parte de la auto-responsabilidad de cada uno para con su vida, respete los ritmos y necesidades del cliente en cada etapa de su desarrollo.

LA TERAPIA GESTALT Y LA PSICOLOGIA TRANSPERSONAL

Ya se ha mencionado cómo la terapia Gestalt pone el énfasis en la presencia plena en el aquí y ahora, compartida por muchas disciplinas espirituales, que ven esta actitud como una forma de evitar las distracciones de la mente y un camino, para el contacto con el Ser interior.

Al igual también que otras disciplinas espirituales, la terapia Gestalt, trata de evitar el pensamiento dual, entendiendo que las dicotomías (imaginación/racionalización, intuición/intelecto, corazón/mente, cielo/tierra, femenino/masculino...) necesitan ser vistas como una delicada danza de equilibrio y armonía, para que abrazando todas las partes de nosotros mismos seamos capaces de comprender la totalidad de nuestro mundo y nuestra inseparabilidad e interdependencia con él.

La salud tiene que ver con permitir a todos los aspectos de nosotros estar ahí, sin juicios; permitirnos desarrollar nuestros sentidos, creatividad, espontaneidad, nuestras emociones. La salud ocurre cuando estamos atentos, absortos y en armonía con nuestra experiencia del momento, y el mundo dentro y fuera de nosotros forma una unidad.

A pesar de que, en ocasiones, por desconocimiento, se tache a la terapia Gestalt de egocéntrica, lo cierto, cuando se conoce, y se practica con profundidad, es que toda la apariencia de individualismo que le da el énfasis que se pone en la escucha de las necesidades, es sólo una parte de un proceso que al final lleva a descubrir el auténtico anhelo del Ser.

La aceptación de la no dualidad, nos acerca a la comprensión de la divinidad y, en mi opinión, si existe un concepto que defina, por sí mismo, lo Transpersonal, éste es el de la no dualidad, criterio que comparte la terapia Gestalt.

LA TERAPIA GESTALT Y EL ECLECTICISMO

Desde la segunda mitad del siglo XX, estamos viviendo una época de eclecticismo, en la que en muchas disciplinas se aúnan herramientas y tendencias diversas; las grandes arquitecturas se convierten en esculturas habitables; las fronteras entre espectáculos musicales, teatrales y *happenings* quedan difusas; muchas pinturas combinan fotografía o diseño por ordenador; y las psicoterapias también rompen sus moldes para dar cabida a nuevas propuestas.

Por suerte estamos siendo capaces de no quedarnos anclados en los conocimientos de nuestra materia, sino que en la medida en que tenemos acceso a información cada vez más amplia, nos vamos permitiendo incorporar nuevos aprendizajes y nuevos conocimientos a nuestra especialidad.

Los años 60 con sus aires de renovación, después de las dos guerras, y la terapia Gestalt, dieron lugar a un modelo de intervención que además de ser una eficaz herramienta terapéutica se convirtió en una filosofía de vida. Surgió en respuesta a las necesidades de los seres humanos en su evolución y, desde este punto de vista, el concepto de terapia no apareció ya forzosamente ligado al de enfermedad, sino que pasó a ser también una herramienta de desarrollo personal y de auto-realización.

Gracias al hecho de que la terapia Gestalt haya traspasado los límites de la intervención clínica, en la actualidad son muchos los profesionales que aplican la terapia Gestalt en sus oficios particulares. Con frecuencia recibo testimonios por parte de ex- alumnos, y colegas, de sus aplicaciones en la docencia, el mundo empresarial, la abogacía, o la mediación laboral.

Al ampliar los ámbitos de aplicación de la terapia Gestalt, ésta se convierte en un instrumento para mejorar el contacto con uno mismo y con el entorno inmediato, sea cual sea éste, y ello puede repercutir en que tengamos unas relaciones mas auténticas y saludables y, por extensión, en que el mundo se haga un poco más confortable.

❖ DEL MARIDAJE ENTRE GESTALT Y ARTE

UN POCO DE HISTORIA

Contextualizando brevemente los orígenes de los creadores de la terapia Gestalt dentro de su momento y lugar, se comprende fácilmente tanto sus vinculaciones con el mundo del arte, como su sesgo contestatario, para la mirada de la época.

Los personajes que se convertirían en piezas claves de la terapia Gestalt, vivieron su infancia en una Europa en que la seguridad y la prosperidad obtenidas gracias a la ciencia y la industria hacía que los intelectuales y artistas, libres de otras preocupaciones que les llegarían después, se centrasen con gran entusiasmo en la doctrina y los métodos de su actividad creadora.

Hacia 1910 la vida artística en Alemania estaba en plena ebullición; existían contactos entre artistas de Francia, Italia, Rusia y Alemania; se celebraban exposiciones conjuntas, y comenzaban a crearse, por diferentes puntos: Berlín, Munich, Zurich, Weimar..., grupos que reunían a intelectuales, personajes del teatro, la literatura, la música, la pintura, la arquitectura, etc.; se habría podido pensar que Europa caminaba hacia su unidad, a juzgar por los intercambios e iniciativas comunes llevadas a cabo por los artistas.

Fritz Perls contaba entonces con 17 años y estaba en contacto con el mundo del teatro y en concreto con el director Max Reinhardt, uno de estos creadores que estaban innovando la metodología de su actividad y que se convertiría con el tiempo en una figura clave del teatro de la primera mitad del siglo XX, además de en una referencia importante para Perls.

La constatación del clima de renovación y creatividad que existía en esta época en Alemania, la tenemos en el nacimiento de la Bauhaus en Weimar en 1919, que agrupaba artistas de diferentes ámbitos con el ideario de crear un arte unitario, totalizador y con la vista puesta no sólo en las élites, sino en el ciudadano común; su ambición era que el arte y la revolución uniesen sus fuerzas.

Desde otra concepción ideológica, pero igualmente de intención renovadora y que aglutinaba artistas de diferentes ámbitos y nacionalidades, apareció el *dadaísmo*.

La catástrofe de la Primera Guerra Mundial, no podía significar para *dadá* más que la caída de la civilización y de las ilusiones cultivadas por los intelectuales y artistas. La pérdida de sentido de sus investigaciones en el terreno artístico, les llevó a una actitud nihilista, de ironía y destrucción de dogmas y convenciones estéticas, que sentían inútiles ante la barbarie de la guerra.

De toda crisis emerge algo nuevo y así en algunos artistas se produjo un desplazamiento revolucionario de su concepción del arte, que dejó de interesarles como pura estética para pasar a ser considerado como un instrumento para la indagación en el ser humano. Nace el Surrealismo.

Es la coincidencia del desencanto por lo poco que había servido tanto derroche intelectual para evitar la masacre, unida a la nueva aportación del psicoanálisis que abría la posibilidad de explorar el inconsciente del hombre, lo que facilitó la aparición del Surrealismo, con todo el mundo de posibilidades que aportaría a la indagación del psiquismo.

El arte estaba presente en los primeros años de la vida de los fundadores de la terapia Gestalt al igual que lo estaba en la culta sociedad alemana de la época.

Fritz Perls, en su juventud interesado por la pintura, ya hemos dicho que se vinculó tempranamente al teatro, mientras comenzaba sus estudios de medicina.

Laura Perls, su mujer desde 1929, se interesaba seriamente por la danza y la música; años después se doctoraría en Psicología de la Percepción.

Paul Goodman intelectual con amplios intereses, dedicaba parte de su tiempo a la literatura lo que sirvió para que años después pusiese en palabras el núcleo fundamental de las teorías que sustentaban la terapia Gestalt.

Todos ellos, y otros que más tarde se incorporarían a su grupo de gestálticos, habían vivido sus años juveniles inmersos en este clima cultural, cuyas inquietudes principales aparecerían después reflejadas, con sus propios matices, en sus formas de entender y aplicar la terapia Gestalt:

— La amalgama de diferentes disciplinas y enfoques, artísticos y conceptuales; como las influencias de Goldstein con su idea del organismo

como un todo indivisible, o las de Lewin sobre la necesidad de cerrar las gestalts, o las ya comentadas de la psicología de la percepción.
- El cuestionamiento de dogmas y convenciones; como el psicoanálisis freudiano.
- La posibilidad de indagar los contenidos inconscientes a través de técnicas artísticas; como el Surrealismo o la dramatización.

Es lógico pensar que ese ambiente de gran inquietud intelectual y de intercambios entre personas de diferentes disciplinas y países, se convertiría en un rico sustrato que daría sus frutos en el futuro. Ya que en aquel momento, la 2.ª Guerra Mundial, lo paraliza todo. En lenguaje gestáltico, podemos decir, que se modifica *el fondo* y que las *figuras* que se movían con facilidad en un clima fértil, son catapultadas en todas direcciones.

En 1933, desaparece totalmente el equipo de la Bauhaus bajo la presión nazi; Perls huye del nacismo a Sudáfrica y muchos de los intelectuales alemanes emigran a E.E.U.U. donde intentan, además de sobrevivir económicamente, seguir con sus investigaciones creativas. Demos un salto sobre ese difícil y confuso período para todos ellos.

Nos situamos en 1946 cuando Perls, con 53 años, llega a Nueva York. Vuelve a contactar con muchos de los intelectuales alemanes instalados allí, y mientras sus colegas, los psicoanalistas neoyorquinos, rechazan su forma de trabajar, es bien recibido por los círculos de la contracultura, comenzando a realizar experiencias terapéuticas con personas provenientes del mundo del teatro, la danza, y la música.

Desde nuestra óptica lejana, parece que hubiese reencontrado el ambiente de sus años juveniles en Alemania, y esta nueva energía se materializa en el libro que escribe junto con Laura Perls y Paul Goodman para recoger los fundamentos de la terapia Gestalt.

Este interés mutuo entre terapeutas gestálticos y artistas ha perdurado en el tiempo, haciendo que muchos terapeutas incorporasen técnicas artísticas en el trabajo con sus clientes, en mayor o menor medida, y que artistas de diferentes ámbitos utilizasen técnicas y actitudes gestálticas.

El primero a citar es Joseph Zinker, además de pintor y escultor, uno de los fundadores del Institute Gestalt de Cleveland, comienza su libro "El proceso creativo en la terapia gestáltica" diciendo: "La creatividad es la expresión de nuestra propia grandeza ..."; en sus talleres grupales, Zinker empleaba frecuentemente la pintura, el movimiento y la música, como inicio para un trabajo posterior de intervenciones verbales.

Michel Katzeff, en los talleres a los que tuve el privilegio de asistir en los últimos años en que vino a vivir a España, además de emplear técnicas de calentamiento habituales en los actores, hacía que los participantes después de exponer ante el grupo su conflicto, lo dramatizasen con ayuda del propio grupo. Sus intervenciones entre tanto iban referidas, en un tono claramente gestáltico, a la toma de conciencia emocional y corporal.

EN LA ACTUALIDAD

También lo han hecho, Thompson-Taupin del Gestalt Institute de S. Francisco, Violet Oaklander, Graciela Cohen, Janie Rhyne, Selma Ciornai, y Pat Allen, entre otros. Todos ellos combinando terapia Gestalt con trabajos artísticos: pintura, poesía, movimiento...

El maridaje continúa; algunos terapeutas gestálticos incluyen trabajos artísticos en sus sesiones clínicas, tanto con niños como con adultos; igualmente algunos artistas del mundo del teatro, de la música o de la pintura, incorporan conceptos de la terapia Gestalt en sus grupos.

Por poner algunos ejemplos, entre los clientes que supervisan conmigo, me he encontrado en ocasiones con:

- Psicoterapeutas que sienten que no avanzan en el proceso con algún cliente que apoyándose en la racionalización, evita entrar en el contacto emocional . Para este tipo de cliente, para el que la palabra es su medio familiar y por tanto seguro, continuar utilizándola pero desde un registro creativo, por medio de cuentos, narrativa o poesía, se convierte en una vía de acceso interior que puede compartir con el terapeuta sin sentirse excesivamente amenazado.

- Profesionales de ayuda a los que la institución para la que trabajan no les permite el abordaje desde la terapia Gestalt de las emociones de sus usuarios, por considerarlo fuera de sus competencias, por lo que se sienten con las manos atadas en su función, han encontrado un camino de actuación a través de trabajos con la fotografía y el video.

- Profesores de artes plásticas en colegios que han sido capaces de resolver conflictos de agresividad en el aula, con propuestas que están a medio camino entre la plástica y la terapia grupal gestáltica.

- Artistas plásticos bloqueados en algún punto de su creación a quienes la terapia gestalt ayuda a descubrir la causa o función de ese *parón creativo*.

- En mediación familiar, cuando las acusaciones cruzadas, y las peticiones de cambios, ya se han repetido tantas veces que no aportan nada nuevo, ponerles a colaborar en un proyecto creativo, en ocasiones, ofrece, con el cambio de registro, una nueva perspectiva.

En la metodología de trabajo con el Arteterapia Humanista que estamos definiendo, el enfoque conceptual se apoya en gran medida en los principios fundamentales de la terapia Gestalt mencionados.

Itinerario del viaje 03

En nuestro viaje a través del Arteterapia usaremos los chakras a modo de estaciones que nos remiten a un lugar geográfico del cuerpo, a recorrer y reconocer, y también a unos productos típicos de la región, que son el temario psíquico adscrito a ese chakra, que vale la pena saborear con detenimiento. En la mochila llevaremos cuadernos de pintura, cámaras fotográficas, músicas variadas, y muchos otros materiales artísticos, pero lo más importante será la disposición de los viajeros para descubrirse en cada recodo del paisaje.

El libro puede leerse apeándose en cualquier estación que al viajero le parezca interesante para la ocasión, pero hay que recordar que lo importante del viaje es el recorrido, que un paso nos lleva al siguiente, que cada uno de esos pasos perfila el anterior y prepara para el próximo. No obstante, es libertad del lector, saltar, correr, o ir a la "pata coja".

ROJO: En la primera estación nos encontramos con que nuestro cuerpo tiene que apoyarse en el suelo, para lo cual necesitamos de la conciencia, del darse cuenta en el aquí y ahora de lo obvio, tanto en lo corporal como en lo relativo a los objetos; para ello vale la pena ayudarse de las tres escuchas, interna, externa e intermedia para diferenciarlas y poder usarlas con discernimiento. Copiando de los árboles su arraigamiento, nos encontraremos con la alerta sensorial y saludaremos a los miedos. Indagaremos en aspectos de la figura paterna que guardamos en nuestro interior y en la búsqueda y hallazgo de *"Mi lugar en el Mundo"*. Nos ayudaremos del dibujo y el collage. Con todo ello estaremos mas cerca de nuestro cuerpo y nuestra realidad, preparados para el siguiente tramo del viaje.

NARANJA: Visitaremos en este lugar la figura materna interiorizada, a ella debemos en gran medida nuestra capacidad de sentir, sin la cual resulta difícil conocer las necesidades, por ello afinaremos la sensibilización corporal y también la de nuestros sentimientos y emociones. Asumir la propia responsabilidad es básico en esta etapa del viaje. Nos vamos a ayudar con el modelado en arcilla, la escritura creativa y el baile, pero sobre todo con la reapropiación de la sensualidad, la sensibilidad y la sexualidad. El vientre y la pelvis serán nuestros aliados y sentirles moviéndose en el baile será una fuente de alegría. Los dibujos de *"El Cuerpo Sentido"*, que no pensado, serán una ayuda indispensable.

AMARILLO: En este paraje nos vamos a encontrar con nuestro ego, al que me gusta llamar con otros apodos: "auto-concepto", "importancia personal"... En fin, sea como sea, nos toparemos con él y con las máscaras que emplea; jugaremos con ellas mediante el *"Teatro de Máscaras"* para quitarles un poco de la importancia que se adjudican. Recorrer en este lugar nuestros mecanismos de defensa, para descubrir cómo nos dificultan en tantas ocasiones la satisfacción de nuestras necesidades, será el premio para esta etapa. Nuestro estómago, bazo, hígado y páncreas, nos lo agradecerán.

VERDE: Estaremos a la mitad del recorrido; entre las estaciones del sur y las del norte. Cruce de caminos, lugar de encuentro para las energías, para el trabajo corporal y el cuerpo poético, para integrar polaridades, para el dar y el recibir. Para tratar de tú a tú a nuestra "sombra" y amigarnos con ella. El desbloqueo torácico y la respiración serán nuestros aliados, y el lenguaje sin palabras de nuestro cuerpo el paisaje a descubrir. Esta etapa tiene un importante reto, la *"Biografía Emocional"*; nada menos que incorporar mis fotos, mi historia, mis otros viajes, a éste, para reconocer las vías muertas, los pasos a nivel y los andenes en los que he visto pasar ya muchos trenes y en los que ya no quiero detenerme más.

CELESTE: El paisaje que atravesaremos tiene como protagonista el cuello, la garganta, el oído y los órganos de fonación. Se dice que la energía de la tierra, pasa por el cuerpo a través de la sexualidad y atravesando el mundo afectivo se une con lo espiritual. Para lograrlo, uno de los controles que tiene que atravesar es el que colocamos en el cuello. Es en ese puesto fronterizo entre la cabeza y el resto del cuerpo donde imponemos nuestro control más férreo y donde la contención de la expresión tiene su amarre. Así que escucha y comunicación, música y voz, sonido y silencio, serán nuestros materiales artísticos en esta etapa.

Entre los compañeros de viaje, después de todo el camino recorrido, habrá ya suficientes guiños y complicidades como para abordar el tema de la confianza en el ejercicio que llamo *"Liberación Cervical Estratigráfica"*. Confieso que le he puesto este nombre tan serio para ayudar a confiar, porque la confianza es la llave sin la cual no se abre esta puerta, y sería una lástima perdérsela, porque ofrece vistas de lo más variado.

INDIGO: Esta estación tiene dos grandes andenes, el hemisferio derecho del cerebro y el izquierdo, cada uno de ellos tiene sus peculiaridades, pero se trata de caminar con un pie en cada andén, hace falta entrenamiento pero es posible, estamos muy bien dotados para este tipo de funambulismo. Desde esa perspectiva contemplaremos nuestro proceso personal, las pautas, los arquetipos por los que nos movemos, haremos "Teatro de Marionetas" con ellos para escucharles con atención y dejarnos sorprender, y disfrutaremos con la *"Dramatización de Sueños"* en este aprendizaje de Inconsciente-Consciente-Inconsciente y tiro porque me toca. Entre esta estación y la siguiente se sitúan los *"Mandalas"* que tanto pueden surgir de la conexión consciente-inconsciente en el terreno del Indigo, como desde una actitud de vacío creativo más relacionado con la pintura meditativa y con una apertura hacia lo espiritual, propia de la siguiente estación.

VIOLETA: Desde la cima de la montaña, se contempla el valle. Desde el simbólico lugar por encima de la coronilla, se admira todo el paisaje interior con perspectiva; sin apegos, o con menos, no seamos ambiciosos. Lo bueno de llegar hasta aquí en este viaje, es que se hará con los pies en el suelo, habitando el propio cuerpo, y sin auto-engaños. Cuanto más anclados estemos a la tierra, más alto, o mejor, más profundo, podrá volar nuestro espíritu. La conexión interna, la respiración y el silencio nos acompañarán a través de la *"Pintura Meditativa"*, de la *"Danza Derviche"* y de los *"Mandalas"* en la búsqueda del vacío creativo y de la conexión con tu Ser Interior.

El viaje no termina aquí, por suerte, ésta es una vuelta de la espiral infinita que es la vida.

Rojo 04

Relativo al Chakra Muladhara

El darse cuenta:

— El cuerpo
— La escucha
— Los miedos

Figura paterna

Mi lugar en el mundo

Ejercicios:

— Dibujo proyectivo
— Lo obvio en los objetos
— Collage

Conclusiones

❖ CONSIDERACIONES RELATIVAS AL CHAKRA MULADHARA

La región corporal a la que afecta es: base de la columna, intestino grueso, plexo sacro; se asocia con los orificios excretores externos: recto, ano, uretra y vagina.

Rige funcionamiento de: Intestinos y Columna vertebral.

Su desequilibrio afecta físicamente a: hemorroides, resfriados, ciática, problemas de próstata, diarreas y estreñimiento.

Se relaciona con el color Rojo, y con el elemento tierra.

Está relacionado con las creencias relativas a la supervivencia: alimento, dinero, casa, desastres naturales, y en general todo aquello que puede dar o restar seguridad, tanto en forma real como imaginaria, y tanto para sí como para su grupo, familia...

La idea de "la familia ha de mantenerse unida a toda costa" puede llevar a la negación de abusos, malos tratos, adicciones, corrupción, etc., en beneficio del buen nombre de la familia.

Tiene que ver con la independencia y con la capacidad de cambio o de adaptación a situaciones nuevas.

El derecho que subyace a este centro es el de "estar aquí", tener lo necesario para sobrevivir, ya sean bienes materiales, amor, o tiempo para uno mismo.

Está vinculado con el instinto básico de supervivencia, y por tanto con el miedo, y con la reacción de lucha o huida.

Las dos funciones asimilación y excreción, que psicológicamente representan tomar lo necesario para vivir y la eliminación de los sobrantes innecesarios, requieren de equilibrio. Representan una disfunción, tanto la dificultad para deshacerse de lo que estorba como el deshacerse con precipitación (desarreglos que cursan con diarreas) de aquello que aún no se ha asimilado adecuadamente (por miedo, muchas veces).

Su desequilibrio afecta emocional y psicológicamente en forma de inestabilidad, depresión, rebeldía, sumisión, sensación de vivir como bajo la ley de la selva, disfunciones sexuales.

La cualidad que se intenta potenciar trabajando desde este centro de energía son la seguridad y el equilibrio.

❖ ARRAIGAMIENTO

Comenzar este viaje de desarrollo personal por los conceptos relacionados con el chakra base, tiene todo el sentido de construir desde los fundamentos, de abajo hacia arriba, y apoyándose bien en la tierra para que nuestra construcción se sostenga.

Experimentamos el mundo a través de las impresiones que llegan a nuestros sentidos; por ello, es indispensable comenzar por atender a nuestra realidad más próxima, el cuerpo; cuanto más vivo esté, con más sutileza e intensidad se percibirá la realidad y más coherente será la respuesta de nuestro organismo.

Dice Lowen: *La "salud mental" es el estado donde coinciden imagen y realidad. La persona sana tiene una imagen de sí misma que concuerda con el aspecto de su cuerpo y con el modo en que su cuerpo siente.* Y un poco más adelante añade: *En el proceso de la curación natural, el primer obstáculo es el desconocimiento que tiene el paciente de las tensiones de su cuerpo.* Esta es una de las tareas que abordaremos a lo largo de todo el proceso, según vayamos recorriendo nuestro mapa corporal.

Comenzaremos por prestar atención a la zona corporal relacionada con este chakra.

> *Manteniéndonos simplemente de pie durante un tiempo, podemos observar cómo es el apoyo en nuestros pies, qué zona de la planta del pie recibe más peso, si nuestras rodillas están o no bloqueadas, si nuestra pelvis se dirige hacia delante, hacia atrás o está perpendicular al suelo, cómo están nuestros glúteos y genitales.*
>
> *Iniciamos nuestro diario gráfico recogiendo estas informaciones mediante sencillos dibujos, que nos hablan de cómo es nuestro apoyo en el suelo.*

Iremos intercalando a lo largo del trabajo con este chakra ejercicios para mejorar el arraigamiento. Comenzamos con un buen ejercicio de arraigamiento. Nos ponemos de pie.

> *Pies paralelos a la altura de las caderas, rodillas levemente flexionadas, pelvis perpendicular al suelo, profundiza la respiración, e imagina que eres un árbol, puedes abarcar ese árbol con los brazos y sentir las raíces, la savia, la corteza... Después píntalo.*

A continuación se pide que hablen de "su árbol", en presente y primera persona, es la primera puesta en común grupal.

En el plano pedagógico comenzar el curso con esta propuesta pone de manifiesto que el compromiso está en hablar de uno mismo; al mismo tiempo sirve para entender y familiarizarse con la técnica de recoger las proyecciones personales sobre las obras, y también para que el grupo se vaya presentando.

Hay que tener en cuenta que no es fácil para un grupo de personas que se acaban de conocer, hablar de sí mismas, por lo que la fórmula de hablar del árbol en presente y primera persona, a muchas les resulta difícil; enseñar la técnica y aceptar las resistencias que probablemente surjan, es un trabajo para el arteterapeuta que requiere equilibrio; por un lado comprensión y por otro firmeza.

Recojo algunos ejemplos:

- *Soy un frutal que tiene muchas hojas y algunos frutos aún verdes, estoy con otros árboles más grandes que yo....*
- *Soy un viejo castaño con una corteza gruesa y llena de heridas...*
- *He dibujado, me cuesta lo de yo soy, un árbol en invierno, con unas delgadas ramas nevadas...*
- *Soy un pino, normal, de los de la sierra, pero mis raíces son profundísimas, se agarran a las rocas...*

Cuando terminan las intervenciones, se pegan los dibujos de los árboles en una pared de la sala creando un original bosque, como metáfora de la presencia de cada uno en el grupo.

❖ DARSE CUENTA

El siguiente trabajo, se centra en la atención a lo obvio, y en diferenciar lo real de lo imaginado. Trabajando con el DARSE CUENTA.

La fenomenología señaló la importancia de la toma de conciencia de la experiencia inmediata, de lo obvio que acontece, y de eso trata la propuesta.

> *Escoged un objeto personal de los que habéis traído hoy aquí, por casualidad, o quizás porque os suele acompañar; se trata de que lo ofrezcáis al grupo poniéndolo en el centro de la sala; es un gesto simbólico porque luego lo vais a recuperar;*

Todas las personas formando un círculo alrededor contemplan esas donaciones individuales al grupo; después de un tiempo, se les dice que escojan uno de los objetos se lo lleven a su sitio y lo dibujen. Sigue una breve ronda diciendo qué objeto han escogido.

> *Sostened el objeto en las manos con los ojos cerrados. Percataros de detalles como textura, peso, olor... y volved a abrir los ojos para registrar sin juzgar, formas, detalles... y añadidlo al dibujo anterior.*

Se hace una pequeña puesta en común en la que es interesante recoger cómo se enriquece la observación del objeto desde el primer contacto más superficial y más condicionado por la utilidad, es decir, por lo que previamente se piensa del objeto, a lo que han ido descubriendo, con las sucesivas aproximaciones.

> *Siguiendo en contacto con el objeto escogido, observad: qué percibo en mi interior, respiración, tensiones, cuáles son mis sensaciones; dadle un nombre a esas sensaciones; llevadlo al dibujo.*
>
> *A continuación, dejad que vuestra imaginación o memoria, siga el curso que quiera y escribid una pequeña historia sobre ese objeto.*

Poder mirar la realidad con otra mirada, es la base de la creatividad; cuando trabajamos con lo obvio, la intención es diferenciar con claridad lo que es obvio de lo que es fantasía, asociación o recuerdo, no anular ninguna de estas posibilidades.

Las primeras aproximaciones al objeto buscaban identificar lo obvio, y con esta última intervención se pide a la persona que se adueñe de la experiencia al poder mirarla de otra manera. Jugamos con la resignificación de la realidad, darle otro uso a los objetos conocidos, cambiarle la función si así se desea en relación a la mirada subjetiva y no al uso común. Esta es una de las bases de la creatividad, pero trabajando con este chakra y siendo el comienzo de nuestra andadura, es muy importante transmitir que no vamos a olvidar ni a enmascarar en ningún momento, lo obvio de nuestra realidad, que la tendremos siempre presente y clara, aunque abordemos otros territorios.

Más tarde, cuando se hace la exposición de los objetos, reales, pintados y novelados es una buena ocasión para ver las diferentes personalidades del grupo, desde las mas realistas y observadoras de detalles, a las mas imaginativas o expresivas.

Acabamos de explorar las tres Zonas del Darse Cuenta:

1. El darse cuenta de la Zona Externa: el contacto sensorial, lo que en este momento veo, toco, degusto, escucho o huelo.

2. El darse cuenta de la Zona Interna: la capacidad de tomar conciencia de sí, tensiones musculares, sensaciones, sentimientos...

3. El darse cuenta de la Zona Intermedia: incluye toda la actividad mental que va más allá del aquí y ahora, todo lo que tiene que ver con explicar, imaginar, recordar, planificar, etc..

Para terminar, cada persona se dirige al dueño del objeto para devolvérselo, y agradecerle, compartiendo lo que más le ha llamado la atención de su objeto; es una forma de devolverlo con un valor añadido, y así lo registran algunas de las personas que reconocen que les dan detalles en los que nunca se habían fijado.

Es una analogía de lo que ocurre con el trabajo grupal, cuando una persona se incorpora a un grupo aporta algo único, algo que solo su presencia puede dar; cuando se va, lo hace enriquecida por la mirada y las devoluciones que los demás le han aportado.

Os invito a que durante las próximas semanas, realicéis de tanto en tanto, algún dibujo escogiendo un objeto del entorno y dibujándolo en el diario gráfico, haciendo el proceso como lo hemos hecho aquí, para ir practicando las tres escuchas.

SUPERVIVENCIA Y MIEDOS

Retomamos el tema de lo obvio desde la reflexión sobre las necesidades básicas para la supervivencia, casa, comida, salud, dinero. Esto tiene que ver con el chakra base, es como si nos preguntásemos ¿cuál es el terreno que estoy pisando?

Reflexionad sobre vuestra situación en relación con la supervivencia, atendiendo a lo obvio del presente.

Simbolizad esa situación incorporando algún elemento o elementos con collage a la tierra de vuestro árbol.

El collage es una herramienta muy aconsejable en los primeros talleres, cuando para algunas personas, el reto de expresarse plásticamente cuando se juzgan poco hábiles, puede constituir una dificultad añadida no deseada. Pero además, las posibilidades peculiares del collage como lenguaje plástico son variadas y susceptibles de analogías; pueden generar volumen, cubrir zonas, o crear superposición de capas de imágenes.

Ya que hemos tratado la supervivencia, es el momento de abordar, por estar íntimamente relacionado con ella, el tema de los miedos, que además muchas corrientes ubican -relacionándolo con este chakra- en la zona de las rodillas; cuando falla la seguridad de un buen apoyo en el suelo, que nos permita ante algo vivido como una amenaza, actuar en consecuencia, lo que ocurre en el cuerpo es un bloqueo, las rodillas rígidas no permiten ninguno de los dos movimientos básicos, ataque o huída, y quedamos paralizados sin posibilidad de reacción, sufriendo los síntomas característicos de sudoración fría, pulso acelerado, respiración corta y rápida... El miedo en sí, el miedo funcional, es un sistema de alarma biológicamente útil para la supervivencia. Pero tiene mala prensa y en parte es por los miedos disfuncionales, que no tienen que ver con amenazas reales sino imaginadas, creadas en esa zona que decíamos antes intermedia, de fantasías de futuro y recuerdos del pasado.

El miedo es una de las sensaciones internas que suelen aparecer al comenzar una actividad o un grupo nuevo como es vuestro caso, y vale más encararlo que tratar de ocultarlo y sufrir el agobio.

Os propongo que, reuniéndoos en grupos de cuatro, hagáis una lista de vuestros miedos, los que aparecen en vuestra vida cotidiana y los que han surgido al venir al grupo. Pondremos después en común los que tienen que ver con el grupo.

Es bueno dar voz a los miedos grupales que aparecen, porque al hablar de ellos y viendo que son compartidos, pierden mucha de su fuerza; no obstante, después de hablar de los miedos haremos otros ejercicios corporales con el fin de ir aumentando el arraigamiento y con él, la seguridad.

Antes hemos trabajado el arraigamiento estático, ahora lo haremos con movimiento, usaremos los pies con ritmo, confianza y capacidad para enfrentar la vida y afirmarnos en la tierra. Pondré algunas músicas de percusión primitivas que nos ayudarán a contactar con la fuerza y energía de lo orgánico; apoyad bien los pies en el suelo, bajad el centro de gravedad abriendo las piernas ligeramente y flexionando las rodillas, soltad la pelvis y dejaros llevar por el ritmo.

El grupo está entregado al movimiento, muy metido en el ritmo, algunos tienen los ojos cerrados; solo hay tres personas que parece que les cuesta más entrar, parece que se cansan, miran alrededor, miran a algún terapeuta y siguen, aparentemente están más en el pensamiento que en las piernas y los pies.

Cambio a una música más suave para ir parando y les pido que se tumben y que sientan su cuerpo, sin poner nombre, solo sentir.

La alerta sensorial no requiere del pensamiento, no significa ponerle un nombre y un concepto a aquello que se aprecia, es puro sentir, desde el presente sin memoria, como si todo lo visto u oído, gustado o sentido lo fuera por primera vez. La alerta sensorial nos vincula al presente.

Después de un rato, les pido que aquello que están sintiendo y que es su cuerpo en activación, lo simbolicen mediante el collage en su árbol; ese árbol que está recogiendo informaciones de como es su estar desde este chakra.

Aparecen algunos antagonismos expresivos, en el sentido de que lo dicho con la palabra no se corresponde con lo que muestra la imagen; también aparecen discordancias en algunas personas entre lo que dicen y lo que expresa su cuerpo. Las intervenciones terapéuticas en estos casos son de señalamiento.

De las tres personas que aparentaban dificultad con el movimiento, una recoge cansancio por falta de costumbre y dos inhibición y pensamientos de fuga. Respeto las resistencias y les animo a recogerlo también en su collage, ya que es igualmente importante. Anuncio que el próximo ejercicio será tranquilo. De hecho me parece una buena práctica alternar los trabajos para evitar la acomodación; de individual a grupal o por parejas, de movido a estático, de introspectivo a compartido.

❖ FIGURA PATERNA

El chakra base se considera relacionado con la figura paterna, con la relación con las figuras o concepto de autoridad, y con nuestra forma de posicionarnos socialmente, es decir, con el rol social que adoptamos. Vamos a revisar ahora estos temas.

Poneros por parejas y dadle al compañero vuestro cuaderno para que escriba lo que vayáis diciendo. Se trata de contarle cómo recordáis a vuestro padre, o figura paterna si tuvisteis un padre sustituto (hermano mayor, abuelo, tío... que cumpliese esa función) cuando erais niños, las cosas que os gustaba hacer con él, las que no os gustaban, lo que os gustaba de él y lo que no. Después pasáis a recordarle cuando erais adolescentes, compartid las características suyas que os vengan a la memoria, lo que os gustaba, lo que no. Y también de cómo es en el presente. Después intercambiáis roles y cuadernos.

Cuando terminéis, escribid una carta a esa figura paterna, diciéndole los agradecimientos y reproches que tengáis guardados del pasado, para poder limpiar la relación y la memoria con ese padre que guardáis en vuestro interior.

Este ejercicio en su primera parte, sirve para refrescar la memoria con respecto a la figura paterna, al tiempo que compartir este tipo de recuerdos con una persona del grupo, facilita que se vaya creando confianza.

La segunda parte del ejercicio, la referida a la carta, es de carácter íntimo y suele ir acompañada de contacto emocional, por lo que después se ofrece la posibilidad de que compartan cómo se sienten.

Releed y subrayad cuales de esas características compartís con vuestra figura paterna y contra cuales os rebeláis, u os habéis rebelado.

Repasad mentalmente cómo ha sido vuestra relación con él y con otras figuras de autoridad, qué sentíais hacia ellas, ¿admiración?, ¿rebeldía? ¿aceptación? ¿desvalorización? ¿alianza? ¿os sentíais protegidos? ¿tratábais de protegerles?

Esta parte del ejercicio, tiene un carácter de reflexión cognitiva y resulta una primera indagación sobre la instauración de comportamientos y de carácter, por la vía de la imitación o de la rebeldía.

❖ MI LUGAR EN EL MUNDO

Vamos a realizar el trabajo que denomino *Mi Lugar en el Mundo*. Empezaremos por crear simbólicamente diferentes mundos: social, familiar, laboral, e íntimo; os indico alguna forma en que se pueden representar, solo a título orientativo:

— El mundo del éxito social, los que están ahí son vistos y escuchados, tienen un papel relevante socialmente. La tarima o el escenario podrían ser los símbolos, o un gran sol amarillo.

— El mundo familiar, entendiendo familia en un sentido amplio, amigos, colegas; los que están ahí les gusta cuidar de su familia, reunirles, ser importante para ellos. El símbolo podría ser la mesa del comedor, la mesa camilla, o un óvalo de colores cálidos.

— El mundo laboral, lo que les interesa es su profesión, disfrutan con ello, no necesitan del reconocimiento social. El símbolo podría ser el despacho, el laboratorio, la camilla de masaje, un rectángulo verde.

— El mundo íntimo, lo que les interesa es la intimidad, solos o con su pareja, sin mucho ruido exterior. El símbolo podría ser un sillón, unos cascos para oír música, un triángulo azul oscuro.

Formad cuatro grupos, cada uno se ocupa de crear una forma, en papel de color, de aproximadamente un metro, que represente uno de esos mundos y le busca un sitio en las paredes o suelo de la sala.

Cuando los mundos estén ya instalados recoged vuestro collage del árbol como representación vuestra que es, y ubicadlo en aquel mundo en el que penséis que estáis, poniendo atención a, dentro de ese mundo que habéis escogido, dónde lo colocáis, centro, arriba, lateral...

La exposición y visita grupal que sigue en la que cada persona habla de su árbol con todos sus aditamentos, de su elección del mundo o sus dificultades para ubicarse, y de cómo ha vivido la presencia o ausencia de otros árboles, es todo un ejemplo de Arteterapia con un objeto transicional y con mediación de lenguajes simbólicos.

La recogida de este trabajo continúa retomando la figura del padre y unas preguntas:

¿Ese es el mundo que tu padre o figura paterna hubiese escogido para ti? ¿Es el que escogió para él? Si ese es el mundo en el que estás ahora ¿quieres mantenerlo? ¿quieres cambiar algo?

Además de las recogidas individuales es muy interesante observar la dinámica grupal que se crea, ya que a cada cambio de posición de un "árbol" cambia el territorio, lo que lleva aparejado que algún otro "árbol" sienta la necesidad de cambiar de lugar, dando lugar a múltiples interacciones en el grupo.

MICRORRELATOS COLOREADOS

ROJO

Remuevo las brasas, añado más troncos, casi se apaga, cojo el viejo soplillo de esparto que no sé ni cuándo, ni por qué compré, y repito el ancestral gesto de atizar el fuego, y junto con las chispas, comienzan a saltar los recuerdos, todos teñidos de rojo.

Recuerdo aquella lámpara roja de mi mesilla de noche, que algunas veces ayudaba y otras no, y el calorcillo, y el olor..., mmmmm, mejor que el de la magdalena de Proust.

Al baile de las llamas se acopla en mi cabeza el ritmo de tambores africanos, sonido y llamas son uno; las plantas de mis pies recuerdan el tacto áspero de la tierra al ser pisada con fuerza, siento mis muslos tensos, sudorosos, fuertes; llevan mucho rato flexionados dando saltos al ritmo de los tambores, sintiendo el calor del fuego por fuera y el calor de la sangre por dentro.

Naranja 05

Relativo al Chakra Svadhisthana Sensibilización Figura materna El cuerpo sentido y expresado — Sensualidad, sexualidad Conexión con un@ mism@ — Necesidades, sensaciones,.. — Asumir auto-responsabilidad Ejercicios: — Dibujos el cuerpo sentido — Diario gráfico, escucha de la necesidad — Modelado en arcilla — Danza — Escritura

❖ CONSIDERACIONES RELATIVAS AL CHAKRA SVADHISTHANA

El nombre sánscrito de este chakra, svadhisthana significa dulzura, "la dulzura del placer"; otra definición es "estar en el lugar que a cada uno le es propio".

Si el primer chakra representa la conexión con la tierra, el segundo representa la conexión con una misma. Desde este centro creamos vida y nos movemos (no solo nos apoyamos) libremente por el mundo material. Está influenciado por el elemento agua, relacionado con las emociones.

RIGE nuestro derecho a SENTIR. Nuestra cultura con frecuencia nos enseña a reprimir las emociones, a no llorar, a no mostrar enfado; si no se nos permite sentir es difícil que sepamos qué es lo que deseamos o necesitamos; hacia ese conocimiento es a donde apuntamos.

El trabajo emocional es el proceso por el cual recuperamos las sensaciones perdidas, para experimentarlas y resolverlas; no siempre es un proceso placentero, a veces nos hace sentir tristes, enfadados, o atemorizados, pero una vez reconocida y expresada esa emoción, las energías corporales dejan de estar encogidas alrededor de ese dolor y pueden expandirse permitiéndonos sentir alivio y aumentando nuestra capacidad de placer.

Cuando es necesario expresar sentimientos, es necesario un oyente neutral, que se abstenga de enjuiciar, de decirnos si tenemos o no razón o derecho a sentirnos como nos sentimos. Lo que estamos reivindicando es nuestra conciencia de lo que sentimos, y nuestro derecho a expresarlo; nuestro derecho a sentir.

Disfunciones que se suelen asociar con este centro:

— Cuando está energéticamente débil puede haber temor al cambio; también se puede parecer inexpresivo, o tener dificultad para sentir las emociones, el placer, o la sensualidad.

— Cuando está excesivamente cargado o sin capacidad de control, ocurre lo contrario, son las personas muy emotivas que pasan con facilidad de un extremo a otro, gobernadas por sus emociones. También es posible que se dejen influir por las emociones de otras personas.

— Cuando está debilitado busca la aprobación y ayuda de otros; con frecuencia la persona se encuentra cansada porque conseguir cualquier cosa le supone un gran esfuerzo, mientras que si está correctamente activado la persona disfruta con lo que hace y tiene energía suficiente para hacerlo.

— Cuando hay dificultad con este chakra, a menudo la persona no se siente con derecho a la felicidad personal, se sacrifica por mantener la paz, estabilidad y equilibrio para los demás, se olvida de su vida ocupándose de las vidas de otros, buscando, inconscientemente, amor y aprobación.

El chakra naranja representa la conexión con el propio cuerpo; desde ahí está relacionado con el trabajo con la figura materna que es quien nos transmite la capacidad para habitarlo, el derecho a sentir.

Llevando el *maternaje* hasta nuestra época adulta, lo relacionamos con el concepto de responsabilidad, ya que es la auto-responsabilidad adulta la que nos permitirá darnos el derecho a sentir y desde ahí, a conectar con la necesidad auténtica.

Para trabajar con este chakra, utilizaremos en primer lugar la sensibilización corporal.

❖ SENSIBILIZACIÓN Y ACTIVACIÓN CORPORAL

No es nuestra intención al trabajar el cuerpo físico, imponer modelos gimnásticos, sino ayudar a que cada persona observe sus límites y se sienta libre de hacer con ellos lo que quiera; pero para llegar a eso hay que comenzar con desarrollar la capacidad de escucha: dónde está el límite real y dónde sólo hay miedo; dónde un movimiento se queda bloqueado por la condición física de la persona o por simple inhibición. Conseguir que cada persona alcance la plenitud de sus movimientos requiere afinar la escucha corporal para hacerse receptivo a las sensaciones.

Para hacer esto posible, se realizan ejercicios por separado de cada parte del cuerpo como un calentamiento que incluye estiramientos y ejercicios de desbloqueo, con movimientos que se van ampliando progresivamente, al tiempo que se incorpora la atención a la respiración, y se va estimulando en algunos casos la emisión de voz como parte de la exhalación y del movimiento.

Dado que en este módulo vamos a prestar especial atención a la sensibilización corporal que nos deberá acompañar a todo lo largo del proceso, se irán intercalando ejercicios plásticos de escucha corporal. Por ello, después de los ejercicios de calentamiento y desbloqueo, se pide:

> *Tomad ahora vuestro diario gráfico y haced un dibujo del "cuerpo sentido"; no se trata de dibujar lo que sabéis de vuestro cuerpo, lo que sabéis que tenéis, sino lo que habéis sentido que tenéis, zonas frías, calientes, tensas, relajadas, podéis emplear colores para representar las diferentes sensaciones.*

Se hace después una pequeña ronda para poner en común lo más importante que han observado; el objetivo es ir creando el hábito de compartir comenzando por temas que no suelen ser conflictivos; además de que escuchar lo que otras personas del grupo han percibido corporalmente, es un estímulo para afinar la atención en las sucesivas oportunidades que van a tener en el trabajo con este chakra.

❖ LA FIGURA MATERNA

El COMPLEJO MATERNO y el COMPLEJO PATERNO son procesos inconscientes y arquetípicos que no se relacionan forzosamente con las motivaciones conscientes de los padres, sino con nuestra vivencia subjetiva de esas figuras.

Analizábamos en relación con el chakra base, lo relativo a la relación con la autoridad, y con la figura o el complejo paterno. Estudiaremos ahora desde el segundo chakra, el complejo materno, cuyo origen está en la idea de que es la figura materna quien da y controla; al crecer puedo incorporar esta figura a mis propios mandatos o ponerlo en alguien de mi entorno.

Sectores en que el complejo materno actúa con mayor fuerza:

— La relación con el propio cuerpo y desde ahí la capacidad de sensualidad y sexualidad.
— El derecho a sentir y a expresar lo que se siente, relacionado con la capacidad de escucha de las propias necesidades.
– La reapropiación de la figura materna, deviene en auto-responsabilidad por el propio cuidado y satisfacción de necesidades.

Para ayudar al grupo en la recuperación de la memoria de escenas asociadas a la figura materna, haremos una relajación seguida de una visualización guiada, en la que se recorrerán diferentes etapas de la vida.

VISUALIZACIÓN GUIADA

Recuerda como eras entre 20 y 30 años, recuerda también a tu madre o figura materna, si tenías una madre sustituta como abuela, hermana mayor... ¿cómo era hablar con ella?...
Después en la adolescencia ¿qué pasaba entre vosotras?...
Entre 8 y 12 años ¿cómo era, qué hacíais?...
De niñas, entre 3 y 8... permite que te venga algún recuerdo.
Intenta recordarte de bebé ¿te llega algún recuerdo de tu madre o de esa persona que se ocupaba de ti? ¿te viene alguna imagen, o alguna sensación?
Vuelve a recordarte creciendo, cuando empezaste el colegio ¿de qué se ocupaba?...

Recuerda en tu adolescencia, tu cuerpo, tu sexualidad, tus novios ¿qué te decía?...
De adulta ya ¿qué recuerdas de ella? ¿cómo era su vida? ¿cómo era con su pareja? ¿qué te transmitió? ¿qué tienes tú de ella?

Muchas personas contactan con la emoción en esta visualización, pero no para todas es el mismo sentimiento. Lo importante es que se han activado los recuerdos y para darles expresión manteniendo un clima de intimidad, les pido que se pongan por parejas, con alguien que se sientan cómodas.

Una de las personas va a hablar de las cosas que ha recordado, diciéndoselo a la otra, que va a representar a la figura materna.

Este ejercicio de hablar y escuchar por parejas los recuerdos ligados a la figura materna, supone un grado de implicación y confianza mayor que lo que se ha hecho hasta el momento; esto debe de ser tenido en cuenta, por los arteterapeutas, por si surgen resistencias por parte de alguna persona, que dificulten el trabajo al resto del grupo.

Para ir modulando las posibles tensiones emocionales, el siguiente trabajo es una reflexión individual de corte analítico.

Escribid sobre la posible relación entre los mensajes verbales o actitudinales respecto al cuerpo, la sensualidad y la sexualidad que os transmitió la figura materna y el cómo vivís y habéis vivido vuestro cuerpo, sensualidad y sexualidad.

Una puesta en común después, nos orienta sobre cómo está el grupo y qué ha sido lo más destacado de sus recuerdos.

- En este grupo hay varias personas que han sufrido abandono por enfermedad de la madre.
- En un caso de muerte materna, no había sentimiento de abandono por haber tenido una madre sustituta.
- El caso más doloroso, abusos por parte de una cuidadora, que ejercía de madre sustituta.
- Agobios por sobreprotección.
- Manipulaciones varias para que haga lo que la madre desea.
- Bastante generalizada la poca atención hacia la sensualidad y el cuerpo.
- Algunos mensajes contradictorios, corporales y actitudinales, entre lo que la figura materna hacía y lo que le decía que había que hacer.

Para concluir, por el momento, este trabajo con la figura materna, se les pide que escriban una carta; el objetivo es poner palabras y recoger las emociones que se les han podido despertar con los ejercicios y también para integrar los descubrimientos que hayan podido hacer sobre el posible origen de sus comportamientos.

Escribid una carta incluyendo agradecimientos y reproches, lo que queréis aceptar como aprendizaje de vuestras madres y lo que rechazáis. Por supuesto no es para dársela, ya que os recuerdo que es con la madre interior con la que estamos trabajando.

Siguiendo la pedagogía de alternancia de ejercicios, se propone otra vez la sensibilización y escucha corporal, ahora con un ejercicio de movimiento con la ayuda de músicas circulares, sensuales, que ayuden a soltar la pelvis.

De pie, vamos a movernos, a dejar que la música nos lleve al baile, prestando atención principal al vientre, caderas y pelvis.
Después volvéis a hacer un dibujo del cuerpo sentido, observando los cambios con respecto al primero.

Aparecen más signos de tensión que en el primero, en parte por haber más escucha, y en parte producidos por los recuerdos de la infancia.

MODELADO EN ARCILLA

La siguiente propuesta es el modelado en arcilla, símbolo de unión de tierra y agua, de la materia con las emociones; es de gran ayuda para la descarga de emociones, y para explorar la libertad de la creación ya que permite con inmediatez, el juego de destrucción y recreación. Es este aspecto el que nos interesa en este momento del grupo; la intención es que se entienda y se integre, que la importancia de la expresión artística en arteterapia está más en el proceso que en el resultado.

La arcilla es un material muy versátil en el que se pueden volcar toda la gama de emociones; puede ser golpeado, estirado, amasado, acariciado; puede trabajarse más o menos húmedo y hasta seco se pueden realizar acciones sobre él, como lijar o pintar; se puede modelar o dejar la huella de alguna parte de nuestro cuerpo en el barro.

Mi propuesta es que cerréis los ojos, para ir amasando y sintiendo el barro, y que poco a poco dejéis que os vaya llegando una forma que simbolice la relación que teníais de pequeñas con vuestra madre o figura materna. Cuando cada una quiera abre los ojos para terminarla.

Cuando han terminado, se hace una ronda para mostrarla al grupo con una palabra o frase como título.

Ahora, con ese mismo material, vais convirtiendo ese recuerdo de la relación con vuestra madre, en una figura del cuerpo que sentís ahora como propio, sin idealizaciones, tratando de sentir y de aceptar lo que hay.

Se trata de aprovechar la conexión que en este momento tienen con su cuerpo para que contacten con su necesidad principal, y relacionen a nivel cognitivo y simbólico, cuerpo y necesidad; el cuerpo puede ser el que avise de la necesidad y también el que ayude a satisfacerla.

A través de la escucha de ese cuerpo que estáis sintiendo, preguntaros cuál es vuestra necesidad, y vais transformando ese cuerpo sentido que habéis modelado, en la expresión de la necesidad.

La concentración en la tarea y la introspección del grupo en este momento es muy grande; a ello han contribuido tanto los ejercicios de sensibilización corporal anteriores, como el *setting* creado mediante la iluminación, la música, y la actividad continuada del modelado.

La ronda que sigue con la exposición verbal y plástica de la necesidad modelada es muy rica, estos son algunos fragmentos:

B: Necesito aprender a poner límites.
F: Necesito permitirme ser más yo, poder jugar sin juzgarme.
C: Necesito contactar más a menudo con mi cuerpo y escucharme más.
I: Necesito un novio, bueno, o algo parecido.
E: Yo también, algo parecido, creo que tengo olvidada mi sexualidad.

A: Necesito dejar de ocuparme tanto de todos y darme un poco más de tiempo para mí.
M: Necesito resolver mi situación económica de una vez.
S: Dejar de pelearme con mi madre o al menos que deje de perturbarme tanto.
L: Escucharme más y hacer que mi pareja me escuche más.

Muestran sus figuras al grupo compartiendo sus descubrimientos.

En este grupo la restauración narcisista que se produce al mostrar la obra realizada no parece un hecho muy significativo, pero es de tener en cuenta este factor cuando se trata de grupos con muy baja autoestima, como el de personas maltratadas, o grupos de reinserción; mi propia experiencia con las personas sin hogar de la Fundación R.A.I.S. (Red de Apoyo para la Inserción Sociolaboral) en los diez años en los que dirigí talleres de expresión plástica, modelado y arteterapia, es que uno de los beneficios que aportaban los talleres, era esa valoración propia y ajena, a través de la obra realizada, que les ayudaba a que su autoestima fuese restaurándose.

En nuestro grupo el objetivo es diferente, aprovecharemos el contacto con la necesidad, para abrir el tema de la auto-responsabilidad por la satisfacción de las necesidades propias.

❖ RESPONSABILIDAD

La Gestalt recoge una idea importante del existencialismo y del sentido de libertad que esta corriente filosófica defiende: cada persona es responsable de su vida dentro del margen de sus posibilidades.

Aumentar la libertad personal es uno de los fines del Arteterapia Humanista y para ello nos ejercitamos en desarrollar nuestra capacidad de sentir, para descubrir nuestra necesidad, ejercitar la capacidad de elegir aquello que necesitamos, y responsabilizarnos de lo elegido y de aquello a lo que renunciamos. Todo ello es ampliar nuestros márgenes de libertad y auto-responsabilidad.

Asumir la responsabilidad plena de los propios sentimientos, pensamientos y acciones. *La responsabilidad no es un "deber"*, dice Peñarrubia, *es un hecho*. Somos responsables de nosotros mismos ...querámoslo o no ; o como decía Perls. *"los otros no están ahí para llenar nuestras expectativas ni nosotros para llenar las suyas"*.

Hacerme cargo de mí, en lugar de evitarlo ocupándome de los otros. Movilizar mis propios recursos en lugar de lamentar, pedir o exigir a otros lo que puedo resolver por mi cuenta. Desprenderme de lo que ya no necesito, sean ideas, cosas o personas. Estos son otros ejercicios de responsabilidad, que también ejercito cuando dejo de culpar a otros por lo que me ocurre.

Y como no hablar de la responsabilidad por nuestro cuerpo, por nuestra salud y enfermedad. Dice la Dra. Adriana Schnake (1995), médica psiquiatra: "*Si se trata a la enfermedad como una mensajera de salud, se la escucha, se entiende el mensaje, antes de que se hayan producido cambios estructurales de importancia, se retirará una vez cumplido su objetivo*", en definitiva lo que la auto-responsabilidad corporal nos pide es que estemos atentos a nuestros síntomas porque algún mensaje, alguna advertencia nos estamos dando a través suyo. Para lo cual otra vez hay que decir que el paso previo es desarrollar la capacidad de sentir, la escucha interna; sentir nuestro cuerpo, sentir nuestras emociones, y responsabilizarnos de ellas, ya que son un producto creado por nosotros.

Continúa diciendo Schnake: "*... Se organizan congresos en los que se reconoce el factor psicológico en el cáncer, en el asma, en la úlcera, y hasta en los accidentes automovilísticos..., pero algo ha pasado que impide al ser humano creer en lo que no ve. (...) Si las enfermedades son nuestro enemigo jamás sabremos si tienen un mensaje positivo para nosotros ¿Quién escucha al enemigo? Pero si nuestro cuerpo está enfermo, nos cuesta pensar; si estamos mental o psicológicamente alterados, el funcionamiento de nuestros órganos se resiente; así que aprendamos cuanto antes a escuchar nuestro cuerpo, a entender su lenguaje y a responsabilizarnos. ¿Cuándo me provoco dolor de cabeza? y ¿para qué me sirve? Cuando quiero parar en mi actividad o no quiero hacer algo ¿a qué recurro?*".

Muchas personas adultas quedan ancladas emocionalmente en una etapa de su niñez, esperando que alguien satisfaga sus necesidades de seguridad y amor que en su momento fueron incompletas.

Nuestra madre interior tiene que saber escuchar cuales son nuestras auténticas necesidades, y tiene que aprender a cuidarnos, esto es, asumir la responsabilidad por una misma.

Se nos ha transmitido una escisión CUERPO-MENTE, falsa y peligrosa para nuestra salud. Antes de continuar ocupándonos de las necesidades y de la responsabilidad, volvemos a realizar el ejercicio de sensar el cuerpo, para tomar el hábito de obtener "diapositivas" de nuestra escucha corporal.

Tumbadas con la espalda en el suelo, respirad dejando caer el peso y a cada exhalación id explorando cómo están los pies, las piernas, la pelvis, hasta recorrer todo el cuerpo; después, como venimos haciendo durante el trabajo con este chakra, haced un rápido bosquejo de vuestras sensaciones en el diario gráfico.

Continuamos esta escucha corporal tumbadas con la espalda en el suelo, piernas flexionadas y ligeramente separadas, plantas de los pies apoyados con fuerza en el suelo, nuca recta, espalda y hombros relajados. Realizaremos un ejercicio que tiene que ver con pelvis y genitales.

Inspirad y exhalad profundamente. Al inspirar subid la pelvis sutilmente, sin despegar las lumbares del suelo, y al exhalar bajad la pelvis; es un movimiento suave acompañado por la respiración, disfrutadlo y cuando os hayáis hecho uno con él, incorporáis la contracción de los genitales al inspirar y soltáis, relajáis al exhalar. Contraigo al inhalar y relajo al exhalar y siento mi vientre y mis genitales.

Este es un ejercicio que a pesar de su suavidad es de mucha activación y muy eficaz para llevar la atención de la mente al cuerpo por la concentración que requiere para hacerlo coordinado.

Era importante esta preparación porque la siguiente propuesta trata de aunar mente y cuerpo a través de la escritura.

ESCRITURA CREATIVA

La escritura es un recurso expresivo, que a muchas personas les resulta familiar, y en nuestra formación insistimos en que se lleve desde el primer momento un diario para ir registrando las vivencias, sueños y reflexiones, pero es frecuente que éste se convierta en una justificación o una demostración y no tanto en un medio de exploración; para este fin suelen ser más eficaces las producciones en que se juega con la ficción, en que los personajes representan al autor, mostrándole de una manera más discreta, por lo que éste, el autor, se permite mas facilmente el juego y el desvelamiento.

Escribir sobre la necesidad-responsabilidad, desde el registro literario que se desee, cuento, poesía, relato, entremés, fábula...;

Algunos de estos escritos trataban humorísticamente sus hallazgos sobre:

La necesidad de sentirse importante, deseada:

Fábula de la ratita presumida, que por querer comer de los diferentes quesos que le traían para conquistarla todos los ratones, sin quererse comprometer con ninguno para poder seguirles teniendo a todos pendientes de ella, finalmente se puso tan obesa que salieron huyendo.

La necesidad de estar en el aquí y ahora:

Cuento de extraterrestres, que devuelven a una persona a la tierra porque no es capaz de darles ninguna información concreta sobre el lugar que había habitado durante 40 años.

La necesidad de parar su hiperactividad, sexual y no sexual, para poder escuchar su necesidad auténtica:

Relato breve seudo-histórico, fechado en el medievo, contando las venturas y desventuras de un soldado huido y sus encuentros con pastorcillas, frailes, y abadesas.

En muchos de ellos se hablaba de su necesidad insatisfecha o ni tan siquiera descubierta, y las formas erróneas de intentar satisfacerla, haciéndose cargo de su responsabilidad en la no consecución del objetivo.

Termina este trabajo con una integración desde lo corporal por medio del baile, con propuestas de diferentes corrientes que tienen como objetivo común el cuerpo global y la unidad psicocorporal, además del disfrute y la expresión.

Para terminar os propongo que os sintonicéis corporalmente a través de la música y el baile, y que vuestro cuerpo vaya asumiendo la identidad del personaje protagonista de la creación literaria que habéis hecho.
Poco a poco, id permitiendo que a través del movimiento, el personaje vaya transformándose y acercándose a la satisfacción de la necesidad.

A continuación, me permito incluir un fragmento de un artículo que escribí para recoger una experiencia de arteterapia a través del modelado en arci-

lla, para ilustrar la capacidad de resonancia interna que puede tener este tipo de trabajo. (publicado en la revista Arte, Individuo y Sociedad n.° 12, 2000, Univ. Complutense)

(...) ES UNA TARDE CUALQUIERA

Falta poco para las tres y media cuando llego al Rincón del Encuentro de la Fundación R.A.I.S., alrededor de doce personas están aquí hoy, algunas leen, otras sestean, alguien apartado en una mesa parece estudiar; saludo brevemente y mientras preparo el espacio para el Taller de Modelado escucho algún comentario, retazos de conversaciones extendidas en el tiempo.

Alguien que sesteaba, llamémosle P., viene, se sienta a mi lado en la mesa, le ofrezco un pedazo de arcilla lista para modelar, pero él prefiere un día más, retomar su pieza con la que lleva trabajando ya varias semanas; es una pieza pequeña cuidadosamente formada y pulida, que algunos considerarían sobradamente terminada pero a la que P. quiere sacar aún acordes más sutiles.

Mientras sus manos iban retocando suavemente la arcilla, sintiéndola, acariciándola, los pensamientos de P. comienzan a tomar voz. Habla de cómo el contacto con la materia le hace a uno sentirse más pequeño, indefenso, abrumado, y que por eso le gusta trabajar con un objeto que siente que no le excede, que puede manejar con un cierto control. A intervalos, silencios, mientras la arcilla baila una danza de amor entre sus dedos. Me viene el recuerdo de algunas piezas de cerámica japonesa, de apariencia sencilla y exquisito acabado, y nuestra charla va derivando hacia el arte zen y cómo el espíritu de la persona se transmite a su obra.

Seguramente, P. no sabe, que el trabajo con arcilla puede ser un poderoso medio para estimular experiencias sensoriales primarias, o que ayuda a la regresión a las etapas iniciales del cuerpo egóico, en las que se desarrolla la simbolización, pero esto carece de importancia, porque él lo está ya vivenciando.

Dos personas más se incorporan al taller, se sientan frente a nosotros, y comienzan a amasar y modelar el barro, hablan de nacionalidades y músicas, la charla ahora se hace expansiva, el tono de voz más alto, la sincronía, una vez más, tiene lugar, y los gestos e impresión sobre la arcilla adquiere toques más enérgicos. Ese clima que ha generado ahora nuestro pequeño grupo, hace que otras personas se aproximen.

P. "Santo Grial",
arcilla modelada y pintada,
aprox. 10 cms.

Mientras, P. silencioso, recogido en sí mismo, mantiene en el hueco entre sus dos manos su pequeña gran obra de arcilla, como si se tratase de un pájaro caído tempranamente del nido, al que es necesario proteger.

Internamente, le doy las gracias a P. por haber compartido conmigo, por un momento, su rico mundo interior.

P. es un hombre de 42 años, con antecedentes de internamiento psiquiátrico, con tendencias depresivas, actualmente medicado y que generalmente se mantiene apartado sin comunicarse con la gente, aunque en ocasiones, mantiene cortos y lúcidos intercambios verbales con las pocas personas con las que establece relación.

Debo decir, además, que una de las dificultades con las que se enfrentan muchas de las personas que acuden a estos talleres, es la de mantener la atención sobre una tarea, objeto u objetivo durante un período continuado, esto forma parte de su problemática vivencial y es también uno de los objetivos ayudarles a desarrollar esa capacidad. (...)"

MICRORRELATOS COLOREADOS

NARANJA

De mi derecha viene el sonido de un goteo desigual, la cueva está solo iluminada por una lámpara de queroseno, oigo un goteo lento más próximo y otro más imperativo hacia el fondo. También desde detrás me llega el sonido de un agua más ronca como cayendo sobre una roca, una sinfonía digna de John Cage.

Me deslizo siguiendo el bajo sostenido del goteo más ronco, y mis ojos comienzan a adivinar escenas de cacerías grabadas en la piedra, me paseo por allí imaginando al anciano de 30 años que seguramente rascaba la pared tumbado en aquel saliente de la roca.

Camino ahora por un corredor más estrecho, la mano siempre por delante de la cabeza "por si las estalactitas", y llego a un pequeño ensanchamiento, el paritorio, aquí, a medio camino entre donde se desarrolla la vida cotidiana y donde tienen lugar las ceremonias sagradas, las mujeres se retiraban para soltar lo que tanto les pesaba ya en el vientre.

Esta cueva huele a vida, a sangre menstrual, a partos sin anestesia, a esperma de mis abuelos, a una vida muy dura donde la creatividad se ponía a prueba cada día, y el premio era la supervivencia.

Amarillo 06

Relativo al Chakra Manipura

Mecanismos de Defensa

— Introyección
— Proyección
— Confluencia
— Retroflexión
— Deflexión

Ciclo Satisfacción Necesidades

Ejercicios

— Máscaras
  - Caricatura
  - Reciclaje
  - Sesión Fotográfica
  - Juego teatral

❖ CONSIDERACIONES RELATIVAS AL CHAKRA MANIPURA

Se sitúa en la región umbilical, abarcando el Plexo Solar, entre la base del esternón y el ombligo. Afecta a los órganos del Bazo, Estómago, Hígado y Vesícula Biliar.

Se le relaciona con el color AMARILLO.

Tiene que ver psicológicamente con la conciencia racional del ego, de la personalidad; también con nuestra necesidad de encontrar la seguridad en la disección racional y lineal de los acontecimientos, lo que algunos denominan la pequeña mente, para diferenciarla de la Mente, con mayúscula, que incluye el concepto global no sujeto a cronologías y que abarca el sentimiento y la intuición.

El elemento que le rige es el FUEGO, que transforma la materia (tierra y agua de los dos chakras inferiores) convirtiéndola en energía y enviándola hacia los chakras superiores.

Esta capacidad de transformación tiene que ver en lo psicológico con LA VOLUNTAD, con el poder de actuar, de obrar con eficacia.

La exploración y potenciación del tercer chakra tiene que ver con el reconocimiento y aceptación de nuestras máscaras y de nuestra sombra, así como con asumir la auto-responsabilidad de nuestra vida, entendiendo ésta como el producto de nuestra propia voluntad.

EXCESO – Afán de dominio. Cóleras, úlcera de estómago.
DEFECTO – Timidez, fatiga, sumisión, trastornos digestivos.

El trabajo con el chakra tercero tiene que ver con abandonar el papel de víctimas, del destino o de fuerzas externas, y asumir el poder personal que nos hace co-creadores de la propia vida, a través del ejercicio de nuestra voluntad.

Es ese ejercicio de la voluntad el que necesitamos para "practicar el acecho" como diría Castaneda (1994), es decir, para estar atentos a cómo y cuándo ponemos en juego mecanismos de defensa que en su momento nos fueron útiles para sobrevivir y que después hemos convertido en automatismos que en muchas ocasiones nos dificultan la existencia.

Las perturbaciones en este centro de energía tienen que ver con nuestra auto-imagen, y es en función de ésta sobre la que construímos nuestros mecanismos de defensa. Por ejemplo, si me he creado la idea de que SOY COMPRENSIVA y NO BELIGERANTE, me armaré una serie de mecanismos que me faciliten cumplir esa auto-expectativa, proyectaré mi incomprensión en otros, no me enteraré de cuándo me agreden para no tener que responder, y si tengo mucha rabia me pelearé con ella internamente, quizás me dañaré sin darme cuenta, pero no la pondré fuera, de esa forma podré seguir manteniendo mi idea de que no soy peleona y de que, a diferencia del resto de la humanidad, yo soy de lo más pacífica.

❖ MECANISMOS DE DEFENSA

Claudio Naranjo (1990) propone mantener el término Mecanismo de Defensa, aludiendo tanto a las perturbaciones de la conciencia (que proponía Freud) como a las perturbaciones de la autorregulación (a las que se refería Perls).

Se nombren de una forma u otra, podemos afirmar que en Gestalt estos mecanismos se consideran como:

- interrupción de la autorregulación organísmica
- formas patológicas de contacto/retirada

En Gestalt la salud se entiende como el ritmo óptimo de la persona entre contactar y retirarse de su entorno. La neurosis es la interrupción o bloqueo de este ritmo. Tan patológicas son las dificultades para contactar como las de retirarse. Tomando este aspecto podemos agrupar los mecanismos del siguiente modo:

CONTACTO	RETIRADA
Introyección	Proyección
Confluencia	Retroflexión

¿CUÁL ES EL PROBLEMA QUE PLANTEAN ESTOS MECANISMOS? En una situación de conflicto de necesidades el individuo tiene que tomar una decisión precisa, se queda en contacto o se retira, sacrifica la necesidad menos apremiante por la más apremiante y eso es todo. Ni él ni el ambiente sufren consecuencias severas. Pero cuando no puede discriminar, cuando no puede tomar una decisión, cuando no está satisfecho con la decisión que ha tomado, ni hace buen contacto, ni se retrae adecuadamente, ambos -individuo y ambiente- se ven afectados.

¿CÓMO TRABAJAR CON LOS MECANISMOS DE DEFENSA? Una de las formas en que se puede trabajar para descubrir estos Mecanismos en Arteterapia es a través de la elaboración y actuación de MÁSCARAS; iremos viendo cómo tratar con cada uno de ellos.

INTROYECCION

Consiste en incorporar, sin "digestión" previa, lo que viene de fuera.

El camino para deshacer la introyección consiste en establecer dentro de la persona un sentido de elecciones propias, y una capacidad para diferenciar el "yo" del "tú". Es descargarse, por tanto, de conceptos ajenos que deben ser expulsados, aunque con el tiempo hayan llegado a sentirse como propios.

Para trabajar con este mecanismo, es adecuada cualquier experiencia que intensifique el sentido del "yo", y que permita discriminar lo propio de lo ajeno.

Cuando trabajamos con las figuras paterna y materna, ya aparecieron, aunque sin nombrarlo, muchos introyectos.

Elaborad una lista con esas frases que nos han acompañado en la infancia, que comienzan con "hay que...", "debería...", "tienes que..." y quizá otras que sin estar formuladas como mandatos, nos transmitieron una actitud elogiosa o crítica hacia determinados comportamientos. Identificad después, aquellos que están presentes aún.

Se ponen en común los introyectos que han detectado, lo que sirve para que muchos reconozcan o recuerden otros olvidados y quizás por eso muy presentes en sus vidas.

La siguiente propuesta es que realicéis la máscara que os obliga a llevar esos introyectos.

Máscaras

La máscara es un elemento ritual tradicional, tenía la función doble de ocultarse y de cambiar de personaje. En Arteterapia, por el contrario, se utiliza para descubrir nuestros diferentes personajes; al mostrarlos en la máscara, les quitamos su poder auto-hipnótico; al ponernos las máscaras elaboradas con conciencia, paradójicamente estamos quitándonoslas, porque mostrándolas al resto del grupo estamos desvelando el secreto.

Existen muchas formas de trabajar y también de construir artesanalmente las máscaras, pero en esta ocasión vamos a utilizar la más sencilla, construyéndola a partir de una hoja de cartón agujereada a nivel de los ojos y orificios nasales, a la que agregaremos pintura y otros elementos.

Una vez realizadas, ponedles un nombre y elaborad un monólogo para presentarlas ante el grupo.

Estos son algunos comienzos de las presentaciones:

F: Los mayores saben más que yo...

A: Soy Inútil. No valgo para nada...
C: Oír, ver y callar es lo mejor que se puede hacer para no tener problemas ...
L: Soy femenina porque soy discreta ...

PROYECCIÓN

Es el proceso inverso a la introyección. Se pone "fuera", en los demás, lo que no aceptamos de nosotros mismos

La piedra angular del proceso de elaboración de la proyección es devolver a la persona los fragmentos de su identidad dispersa. Se trata de tomar la responsabilidad de nuestra propia vida, haciéndonos cargo de aquellos asuntos que, a modo de espejo, nos reflejan los demás.

Algunas formas de trabajar con el mecanismo es preguntarnos si aquello que digo de los demás tiene algo que ver conmigo, o simplemente ponerlo en primera persona y ver si nos resuena.

Piensa en alguien que no soportas, con quien te cuesta trabajo relacionarte.
Realiza una caricatura de esa persona señalando esos rasgos que te la hacen incómoda o desagradable.
Date cuenta desde que oyes la consigna, mientras lo haces, y cuando lo terminas, cuales son los introyectos que aparecen y sigue con la caricatura.

Después, por tríos, ayudaros con estas preguntas:

¿Cuáles son los introyectos que me han venido con la propuesta, ejecución y terminación de la propuesta de dibujo de caricatura?

¿Por qué no soporto a esta persona?

¿Cuál es el introyecto que hay detrás? Ej. si no le soporto porque es un vago, el introyecto podría ser: "tengo que ser trabajador".

¿Qué hago yo con eso? (no lo acepto en mí ni en nadie, o no me lo permito aunque lo desee...).

Cuando tengáis clara la proyección, por ejemplo, soy un vago vocacional, pero como no me lo permito no lo soporto en nadie, realizáis una máscara, esta vez con elementos reciclados, pudiendo incorporar diferentes proyecciones.

La utilización de material de reciclaje es en sí misma enriquecedora, ya que exige ver una realidad próxima, un objeto concreto, con una mirada creativa, que posibilita otras opciones de uso; esta capacidad de contemplar diferentes opciones para resolver un problema, es la base de la creatividad.

Este tipo de máscara requiere de más elaboración, como lo requiere así mismo, el identificar y recoger las proyecciones, es por esto que se elige dedicar más tiempo a este mecanismo de defensa.

También se da importancia a su presentación en el grupo por medio de la sesión fotográfica:

Cuando tengáis la máscara de las proyecciones preparada, vais a salir al centro del grupo a presentarla, sin palabras, solo con lenguaje corporal, llevando al cuerpo la actitud que implica esa máscara.

Se les pide que para que no se desconecten mediante la sobre-actuación, traten de llevar la atención al cuerpo, que intenten percibir cómo se sienten corporalmente cuando utilizan esa máscara; mientras tanto se les avisa que van a ser fotografiados.

Fotografía

Aquí tenemos uno de los usos de la fotografía como herramienta terapéutica; la fotografía normalmente lo que aporta es el juego con el espacio y el tiempo. Pero en este caso, la fotografía es utilizada como soporte de una relación entre el fotógrafo y los espectadores de un lado, y la persona fotografiada del otro, que pone de manifiesto lo que quiere mostrar de sí misma, asumiendo su identidad o disfrazándola a su manera. Así que, en este caso la fotografía hace visible la mirada de los otros, y también es una memoria gráfica de su actitud corporal.

CONFLUENCIA

Cuando una persona es incapaz de diferenciar el límite entre él mismo y el ambiente se dice que está en confluencia. Normalmente se emplea este mecanismo de evitación del contacto real para huir del peligro de confrontación o de diferenciación.

Para corregir esta tendencia, la persona ha de experimentar las elecciones, necesidades y sentimientos que son sólo suyos y que no tienen, necesariamente, que coincidir con los de los otros. En definitiva, atender a cuáles son sus necesidades, diferenciarse e individualizarse.

Para trabajar el siguiente mecanismo, Confluencia, se les pide que observen en su vida cotidiana, las veces que hablan en plural: nos gusta, preferimos, nos aburríamos,... También las ocasiones en que no se comprometen con una opinión o elección, y dicen: como quieras, me da igual, decide tú...

> *Por tríos, explorad de qué forma, con quién, cuándo usáis este mecanismo.*

Cuando hacen la recogida de estas observaciones algunas personas están sorprendidas por haberse dado cuenta de lo habitual que es en ellas este comportamiento; por el contrario, otras apenas lo registran.

Es un mecanismo que una vez que se conoce es fácil de entender y detectar por lo que no es necesario insistir sobre él; otra cosa es conseguir desprenderse de este comportamiento cuando se ha hecho un hábito. Hacerlo requiere mantener una atención constante sobre la propia necesidad.

RETROFLEXIÓN

Consiste en delimitar excesivamente el límite entre yo y el entorno, así que en vez de hacer contacto traspasando ese límite, la persona se vuelve hacía sí y se hace a sí mismo lo que le gustaría hacer a los demás.

El movimiento hacia el desarrollo consistiría en redirigir la energía de modo que en vez de limitarse a operar dentro del individuo, quedara en libertad para iniciar una relación con algo exterior al yo, por lo que conviene que el sujeto tienda al contacto, pero frecuentemente hay que deshacer previamente la lucha interna. Examinar actitudes, gestos, ademanes... que podemos señalar, para que la persona se haga consciente de su interrupción del contacto hacia el exterior.

Por subgrupos van a ayudarse en la observación del mecanismo de Retroflexión, inculpándose cuando no procede, lastimándose cuando se está irritado contra alguien, autocriticándose o creándose algún tipo de malestar físico, como dolores de cabeza o de estómago, con ocasión de enfados no expresados. Esto lo van a hacer mientras realizan las siguientes máscaras.

> *Vais a trabajar sobre la máscara como espejo que os devuelve vuestra imagen según la ven otros. Formar grupos de cinco personas, una de las cuales se ofrece para ser la primera a la que le hagan una máscara entre todas las demás. Se trata de ir simbolizando en la máscara cuáles son los mecanismos que apreciáis en ella al tiempo que se lo decís, concretando la ocasión en que lo habéis observado. Este trabajo exige un ejercicio de sinceridad y también de atención, ya que durante toda la ejecución vais a ir observando, además de los otros mecanismos que ya se han explicado, cuándo alguien de vuestro grupo da alguna señal de estar utilizando el mecanismo de la Retroflexión . Podéis emplear para hacer la máscara, dibujo, pintura y collage. Acordaros de poner palabras a lo que estáis representando.*

Este trabajo supone un paso adelante en la dinámica grupal, ya que se les pide que se den devoluciones unos a otros, y da pie a una discusión abierta en cada subgrupo, en donde se observan en vivo, los mecanismos de defensa más habituales: quién se une siempre a la opinión de la mayoría, quién se distrae, quién ve aquello que después le devuelven a él...

Después de la realización de estas últimas máscaras, cuando el cansancio hace mella en el grupo, es un buen momento para frustrar el deseo de descanso, proponiendo el trabajo con el mecanismo de deflexión, de esta manera será fácil detectarlo.

DEFLEXION

En palabras de los Polster, "la deflexión es una maniobra tendente a soslayar el contacto directo con otra persona, un medio de enfriar el contacto real".

El deflector sí toma contacto, pero se desensibiliza para evitar las emociones, desplazando la energía hacia otra cosa. Algunas formas de deflexión son el humor o la ironía, la verborrea, la somnolencia, ser diplomático en lugar de franco, y una forma muy frecuente es la racionalización, que algunos autores, para darle mayor relevancia, nombran como un mecanismo diferenciado.

Se pide elaborar otra máscara poniendo a disposición del grupo gran cantidad de materiales, papeles para collage, pinturas, textiles, materiales de reciclaje y de manualidades, para la realización de la máscara que represente sus formas de deflectar.

> *Disponéis ahora de todos estos materiales variados para que hagáis una máscara, en la que expreséis vuestras formas favoritas de Deflectar; os sugiero que observéis vuestras posibles deflexiones durante su ejecución.*

No hace falta esperar mucho para poder observar en vivo y en directo cuál es su peculiar manera de deflectar en este momento.

> *¿Qué? ¿qué has dicho? perdona, estaba distraída.*

Tienen ahora una buena colección de máscaras, y una primera aproximación a lo que son sus Mecanismos de Defensa; es el momento de digerir, observar en la cotidianeidad, y re-elaborar. Se les pide que tomen la colección que tienen como bocetos para que, a partir de ellos y de lo que vayan observando de su comportamiento durante el mes, realicen al menos dos máscaras sobre sus mecanismos más habituales, para trabajar con ellas en el próximo encuentro.

> *Llevaros todas vuestras máscaras y consideradlas como bocetos, con ellos y la auto-observación durante el mes, elaboráis al menos dos máscaras sobre vuestros mecanismos cotidianos.*

Se les ofrece información sobre la realización de máscaras en papel maché, pero también se les indica que pueden hacerlas empleando como base las que venden o utilizando las técnicas de collage o de material reciclado que hemos empleado hasta ahora. Lo importante del trabajo con máscaras en Arteterapia, es todo el proceso de identificación con la propia máscara, ya sea a través de la idea como de la fabricación, así como también el potencial posterior que tiene de distanciamiento y observación.

Por otra parte, no es casual el efecto de "saturación de las propias máscaras", es una consecuencia buscada, que lleva al cuestionamiento de ¿y con tanta máscara quién soy yo?

❖ CICLO DE SATISFACCIÓN DE NECESIDADES

En el siguiente encuentro se explica El Ciclo de Satisfacción de Necesidades, que es una herramienta fundamental en la psicoterapia gestalt, y que también podemos encontrar denominado como ciclo de autorregulación organísmica, ciclo de contacto-retirada, o ciclo gestáltico.

La idea del Ciclo como tal, parte de Perls influenciado por Reich y su ciclo de 4 tiempos: "tensión, carga, descarga y reposo".

Se trata de la descripción, paso a paso, del proceso desde que surge una necesidad hasta su satisfacción o la interrupción del proceso con la consecuencia de que dicha necesidad queda sin cubrir, lo que denominamos una "gestalt inconclusa" que tiende a repetirse en su búsqueda de satisfacción.

Las fases del Ciclo:

1. Del REPOSO inicial...
2. ... aparece una SENSACIÓN.
3. Cuando ésta se hace clara y tomamos CONCIENCIA, le damos nombre.
4. Se moviliza en nuestro interior una energía, ENERGIZACIÓN o excitación corporal.
5. Se traduce en una ACCIÓN.
6. Que nos lleva a satisfacer nuestra necesidad o a entrar en CONTACTO con el entorno para satisfacerla.
7. Una vez satisfecha, se produce la RETIRADA hasta que emerja un nuevo ciclo.

   Se les pide que se concentren en una necesidad actual insatisfecha y que, siguiendo paso por paso el Ciclo explicado, ubiquen en qué momento y con cuál de sus máscaras se impiden alcanzar la satisfacción de la necesidad.

Los Mecanismos de Defensa representados en las máscaras, se establecieron en su momento porque resultaban útiles para sobrevivir; lo que nos proponemos ahora, es saber si en el momento presente, continúan siendo útiles o si se han convertido en un lastre que dificulta la satisfacción de las necesidades.

Todo el trabajo realizado hasta este momento relativo a este chakra, tiene un componente mental importante, de darse cuenta desde lo cognitivo, habiéndose utilizado las herramientas artísticas como forma de plasmar, recoger y sintetizar los hallazgos hechos. Ahora invertiremos el proceso, dejando que cuerpo y máscaras hablen, y a ser posible, sorprendan al protagonista.

Nos servimos del JUEGO TEATRAL, apoyándonos en dos recursos: el "fundido en negro" o "folio en blanco" del que habla Lecoq y "la distancia estética", concepto acuñado por Landy.

No hay guión previo, no se trata de hacer comedia. Saben cuál es su necesidad insatisfecha. Se trata de que se sientan dentro de su máscara y partiendo del "fundido en negro" la vivan, la dejen —se dejen ser— para conocer-se. Abordaremos la improvisación por medio de la recreación psicológica silenciosa, tratando de conectar con una situación de ingenuidad primaria, de inocencia y curiosidad, para que, evitando el discurso y lo explicativo, la palabra o la acción que nazca del silencio y de la conexión interior, esté cargada de fuerza y significado.

Se pide que salgan a improvisar con sus máscaras por grupos pequeños, de 3 a 5 personas, según se vayan conectando con la experiencia.

> *Se trata simplemente de pasear por la zona de la sala, delimitada como escenario, en silencio, sintiendo la respiración, con la máscara puesta, en contacto con la necesidad insatisfecha que cada uno ha reconocido y dejarse vivir lo que surja, con la mayor espontaneidad posible.*
>
> *Mientras tanto, si alguno de los espectadores se conecta con lo que esté pasando puede ponerse su máscara y salir al escenario.*
>
> *Igualmente, si alguno de los que están en el escenario desea retirarse, lo hace quitándose la máscara al salir y pasando al grupo de espectadores.*

Landy habla de la "distancia estética" proporcionada por el uso de metáforas o analogías. En nuestro caso esta distancia estética la proporcionan: por un lado el escenario, espacio de la sala delimitado que separa la zona de la representación, de la zona de realidad en el "aquí y ahora" del grupo; por otro lado, la actuación desde el personaje con nombre propio que representa la máscara. Todo ello facilita para el que está actuando, vivir lo emocional y lo racional simultáneamente.

El resto del grupo mientras tanto, en su papel de espectadores, pueden identificarse con lo que observan y relacionarlo con su propia experiencia, es decir, resonar. Así van apareciendo de tanto en tanto como hilos conductores o pequeñas tramas que conectan la secuencia de acciones de una persona con otra, sin que ello signifique que el significado interno deje de ser individual.

Todas las personas del grupo parecen muy conectadas con la acción y con sus máscaras, produciéndose interacciones por parte de algunos. Después hay una ronda grupal en la que se comparten las experiencias que cada uno ha vivido, y es interesante para ellos darse cuenta de que la misma acción, a veces, era vivida diferente por cada una de las personas implicadas.

Algunos ejemplos de los comentarios surgidos:

> *M:* (que muestra una máscara aterradora, trata de conectar con algún otro desde lo tierno, y éste huye asustado) *No tenía conciencia de que mi máscara pudiese dar miedo .*
>
> *E: Mi máscara era de protección, y he visto que me debilita porque no me permite sacar mi fuerza.*
>
> *F: Pretendía que mi máscara hiciese ver lo fuerte que soy, como un reclamo, y he visto que me aísla, aunque algunas personas sí se acercaban como a que las protegiese.*
>
> *C: Quería mostrar serenidad, imparcialidad, y me he dado cuenta de que era más bien como no intervenir, de hecho parecía que nadie me veía.*

El grupo ha realizado un trabajo de auto-indagación aplicando conceptos nuevos, que integra al relacionarlo con su necesidad no satisfecha en el presente. Ha descubierto cuáles son los comportamientos promovidos por los mecanismos de defensa que motivan esta no satisfacción de la necesidad; esto es entendido desde lo racional-cognitivo y completado en unos casos y subrayado en otros, por medio de la exploración corporal intuitiva a través de las máscaras y del juego teatral. La sensación general es la de haber comprendido actuaciones concretas personales que puede serles útil modificar.

MICRORRELATOS COLOREADOS

AMARILLO

Uno de los primeros que llegó a aquella extraña reunión fue el poeta, místico y pintor inglés William Blake.

Poco después aparecieron por allí otros simbolistas esotéricos que se interesaban por las investigaciones metafísicas y metapsíquicas.

También había un personaje que parecía atraer la atención, era Goethe que estaba discutiendo con Newton, sobre el color y la luz.

Una mujer a mi lado les escuchaba lanzándome miradas de tanto en tanto.

Están todos locos, dijo ella, me han puesto dolor de estómago, no soporto a todos estos artistas-filósofos, filósofos-científicos y místicos-poetas, todos tan intelectuales; y muy acalorada continuó criticando y argumentando sobre el Romanticismo-Espiritualista frente al Empirismo-Materialista, hasta que, de pronto, se calló un momento, me miró y me preguntó: ¿será esto a lo que llaman proyectar?

Verde 07

Relativo al Chakra Anahata

El cuerpo como matraz alquímico

Polaridades

Ejercicios

— Sensibilización corporal
— Desbloqueo y Energización
— Cuerpo Poético
— Modelado en arcilla
— Masaje
— Baile

Biografia emocional,

— Previos, Acción y Recogida
— Fotografía

❖ CONSIDERACIONES RELATIVAS AL CHAKRA ANAHATA

Localizado en el centro del tórax, es el centro de conexión entre los tres chakras inferiores (físico-emocionales) y los 3 chakras superiores (mental-espirituales), por lo que representa el lugar de encuentro entre las energías que ascienden de la tierra y las que descienden del cielo, donde ambas se funden y se materializan en el dar y el recibir, y en la fusión de polaridades, izquierda/derecha, arriba/abajo.

El color que lo representa habitualmente es el verde, que es el encuentro de colores fríos (azul) y calientes (amarillo).

Está asociado con el elemento "AIRE". El aire llena todo el espacio en donde penetra, y no está sujeto a la fuerza de la gravedad, de la atracción de la tierra; a diferencia de otros elementos, el aire no es visible, es sutil, por el aire desplegamos las emociones, y mediante el aire, la respiración, las retenemos, bloqueamos, o las dejamos fluir. Dejar de retener la respiración es dejar de retener las emociones.

El aire simboliza la libertad; no saber amar significa ahogarnos y ahogar al objeto amado, por aferrarnos, por miedo a la pérdida. Por tanto, aprender a amar es también aprender a ser libre y a respirar y dejar respirar.

El funcionamiento óptimo de este chakra, nos aporta armonía, equilibrio, ternura, compasión y amor. Su disfunción nos llega con frialdad emocional, insensibilidad, miedo a exponernos, ansiedad, superficialidad. Físicamente, está relacionado con problemas cardíacos y respiratorios, asma, alergia, dolor entre los homoplatos (zona que en alguna tradición denominan "de la traición " (del puñal en la espalda, diríamos nosotros) y dolor de hombros (cargar con pesos o responsabilidades ajenas).

❖ EL CUERPO COMO MATRAZ ALQUÍMICO

En este chakra trabajamos principalmente con el cuerpo, ampliando su potencial: abriendo, movilizando, desbloqueando y aireando, no solamente los pulmones, sino todo el flujo sanguíneo y a través suyo las células del organismo, para descubrir después sus recursos expresivos, emocionales y poéticos.

Después de la auto-observación intensa llevada a cabo durante los trabajos relacionados con el chakra anterior, es conveniente reclamar la atención a lo corporal, y aunque desde cada centro hemos dado un espacio importante a los ejercicios corporales poniendo una especial atención en aquellos relacionados con la zona implicada, al llegar a esta parte del proceso, es indispensable profundizar y afinar la conciencia corporal.

Existen en muchas personas, diferencias importantes entre el cuerpo sentido y el cuerpo imaginado; no se nos educa cuando niños para sentirnos sino para pensar. Ello trae como consecuencia la poca conciencia corporal que tenemos de adultos y que se agrava con los posibles traumas vividos, como falta de contacto físico-afectivo por parte de los padres, maltratos, abusos... siendo éstos los agravantes más habituales. Lo cierto es que todos los adultos, en mayor o menor grado, tenemos dificultad para percibirnos sutil y globalmente.

Siendo como es el cuerpo nuestra principal fuente de información, vale la pena no sólo cuidarlo sino también aprender a escuchar sus mensajes.

Lo haremos mediante el desbloqueo, la sensibilización y poniendo el énfasis en la atención a la respiración. Para ello está bien comenzar con estiramientos y ejercicios de desbloqueo de las articulaciones; seguidos de asanas de Yoga que incorporan ya la respiración de una forma específica y ayudan a ir focalizando la atención en zonas concretas.

Aprovechamos esta activación corporal como desencadenante de creación, pero sobre todo como desencadenante de implicación.

> *Los ejercicios que habéis realizado para ayudaros a tomar conciencia del aquí y ahora, de vuestro cuerpo y el grupo, os pido que los completéis ahora, reflejando en vuestro diario gráfico cómo estáis; utilizad el dibujo como otras veces, con la libertad de expresaros de forma figurativa o abstracta y pudiendo emplear los colores para mostrar las sensaciones y emociones.*

Este primer dibujo sirve para obtener información de cómo está el grupo sin recurrir a la expresión verbal-racional, de esta forma se pretende no interrumpir el proceso de focalizar en lo corporal.

Después de este ejercicio de calentamiento, el grupo ya centrado y corporalmente preparado, está listo para entrar en algunos ejercicios de Bioenergética; según Wilheim Reich cada carácter tiene la energía vital bloqueada, estásica, en alguna zona corporal, por lo que la terapia corporal es un factor importante en el desarrollo psíquico. Los principales bloqueos que se aprecian en los dibujos corporales que acaba de realizar el grupo se centran en la zona de la pelvis, zona torácica y cuello.

Se comienza con ejercicios de arraigamiento para continuar con el desbloqueo de la pelvis, ello da lugar a algunas descargas de rabia contenida, que son apoyadas por los terapeutas para contrarrestar el miedo que suele producir soltar lo retenido durante largo tiempo. Aunque esto se dice en una frase, el trabajo lleva tiempo y dedicación por parte de todo el grupo.

MODELADO EN ARCILLA

La recogida de esta experiencia, siguiendo con el propósito de no racionalizar, se va a hacer mediante la expresión con arcilla, con un enfoque diferente del que empleamos al trabajar el chakra naranja. Recordemos que la arcilla permite desde la descarga emocional a la sensualidad del modelado, y que deja una memoria objetiva de lo sentido y expresado, que después se retomará cuando llegue el momento de analizar, conceptualizar y poner palabras a la experiencia.

Mientras el grupo está con los ojos tapados con un antifaz, se les ofrece un recipiente con arcilla para que sean sus manos las que decidan qué cantidad coger, se les pide que la calienten, la amasen, entren en sintonía con ella, la huelan, se unifiquen con ella, sientan si la quieren más seca o más húmeda, más consistente o más blanda. Y que vayan después permitiendo que las manos prueben gestos diversos, y que el cuerpo vaya acompañando estos gestos; deteniéndose en aquellos que su cuerpo les pide, ampliándolos, repitiéndolos, escuchando qué les dice, probando otros nuevos.

> *Con los ojos tapados como estáis, tomad la cantidad de arcilla que os pidan vuestras manos, que sean ellas quienes dirijan la acción al principio; no se trata de modelar en esta ocasión, sino de expresar diferentes acciones; acariciar, pellizcar, golpear, perforar, amasar, lo que os apetezca, y poco a poco vais incorporando el cuerpo en esa acción, que el movimiento parta de la pelvis, involucre el vientre, la columna vertebral y los brazos. No busquéis una forma, dejad que el cuerpo se exprese.*

La actividad corporal que se emplea en el trabajo con la arcilla en esta forma, desencadena con facilidad respuestas emotivas, resonancias ligadas más a los gestos que al resultado.

Llevan a cabo esta expresión con la arcilla con los ojos tapados con el fin de no distraerse de su escucha corporal y disminuir al máximo el juicio mental. Las emociones que se han podido contemplar son variadas igual que el proceso de ejecución y las obras terminadas:

> *B: Golpea la arcilla con los puños, clava los dedos, y no contenta con eso empieza a tirarla levantando los brazos y arrojándola al suelo. Entrecejo fruncido.*
>
> *L: Repite un gesto de alisar con las yemas de los dedos, humedecer y volver a alisar, parece muy sensual.*
>
> *C: Parece desconcertado al principio, como si no supiese qué hacer; después redondea la masa con las palmas de las manos, con gestos circulares repetidos.*
>
> *F: Estruja y estruja la arcilla añadiéndole agua; el gesto principal parece estar en clavar los dedos apretando el montón de arcilla muy húmeda para que ésta se escape por entre los dedos.*

> *M: Coge arcilla una y otra vez hasta hacer un gran montón que agujerea para llenar de agua que le resbala por todas partes, se tira al suelo con la cabeza próxima a la arcilla y trata de contener el agua añadiendo más y más arcilla mientras sonríe.*
> *E: Modela algo que podría ser un animal, lo hace usando toda la mano, con gestos concretos, perfora, pellizca, añade pedazos.*
> *S: Parece estar haciendo una figura de mujer, va poco a poco, recorriendo a menudo la forma con los dedos.*
> *A: Inicia diferentes acciones como buscando, finalmente se cubre la mano izquierda como haciéndose un guante de arcilla y se emociona.*
> *I: Amasa incorporando en el movimiento todo el cuerpo, está sudando y con una expresión muy concentrada.*

Pasado un tiempo suficiente, se les quitan los antifaces para que vean hacia qué forma les han llevado sus gestos y para que, si lo desean, lo acaben, le den un título y compartan con el grupo lo más destacado que les haya aportado la experiencia.

> *Ahora que ya podéis ver la forma que ha surgido de vuestra expresión, dejad que ella os hable, que os sugiera un título y, si lo deseáis, podéis darle los últimos toques.*

En la ronda que sigue, muchas de las personas del grupo no desean hablar o son muy parcas en sus palabras. Están introspectivas, como digiriendo aún la experiencia que acaban de tener. Por lo que parece oportuno dar un tiempo para que tomen notas en su cuaderno; es otra forma de recoger lo que han vivido que parece más acorde con su momento.

MASAJE

El segmento corporal que se está trabajando es el plexo torácico; el desbloqueo de esta zona suele ir acompañado de una sensación incómoda de no comprender, que quiere decir no controlar, lo que se está viviendo, por lo que es muy importante cuidar el "*setting*", todo el encuadre terapéutico que reasegure: la sala, la estructura del trabajo, las normas, la presencia atenta de los facilitadores; en resumen, todo aquello que hace que la persona, que está desmontando una coraza que tuvo en su momento una función de supervivencia emocional, pueda abordarlo en la confianza de que está en un espacio de seguridad. Por ello mismo, la recogida de estos trabajos se hace desde el *maternaje*, cuando ya hay confianza en el grupo y se ha alcanzado un buen grado de contacto con el propio cuerpo; entonces es el propio grupo, el que cumpliendo la función madre, contiene la experiencia desde el contacto y el masaje estructurante.

Buscad una pareja con la que os sintáis cómodas; quien vaya a recibir el masaje se tumba boca abajo y quien lo vaya a dar que mantenga todo el tiempo posturas que le sean cómodas. Se trata de un masaje relajante, de presión natural, esto quiere decir que lo hagáis a través de vuestro peso y no por la fuerza de las manos. Imaginad que estáis reuniendo a la persona bajo su piel, compactando sus fragmentos; es un masaje amoroso, como el que daríais a un bebé para hacerle sentir vuestra presencia.

La capacidad para dar y recibir forma parte del trabajo con este chakra, y mediante el masaje se evidencia cualquier dificultad en este sentido: tanto quienes se juzgan mientras dan, como quienes les encanta dar pero no encuentran el momento de recibir, quienes les cuesta trabajo concentrarse en el recibir y se distraen con pensamientos que les sacan de la experiencia, y todo un abanico de posibilidades, incluyendo las de quienes lo disfrutan.

Esta actividad de dar y recibir termina con un intercambio verbal entre las parejas compartiendo su experiencia.

BAILE A TRAVÉS DE 7 PAISAJES

Haya pasado un día, una semana, o un mes, después de quitar, o al menos agrietar una coraza muscular, el sujeto, y nunca mejor dicho, ahora menos encorsetado, suele tener la sensación de vulnerabilidad. Acudir al Chi Kung o cualquier técnica de similares características, es una muy buena opción, para ayudar a que la persona se sienta bien afirmada sobre sus pies, con una respiración amplia que la llene de energía y para que se sienta habitante y dueña de su cuerpo.

Después de estos ejercicios preparatorios, os propongo continuar con el movimiento a través del baile; voy a ir poniendo diferentes músicas y, en la medida de lo posible, os invito a que el movimiento no parta de una idea previa sino que dejéis que la música os mueva.

El objetivo es poder experimentar, desde lo corporal, los posibles cambios que haya habido e irlos integrando. La propuesta es una experiencia de baile libre desde la escucha y expresión de diferentes tipos de música, que conllevan su paralelo emocional permitiendo su plasmación corporal; el recorrido va desde los ritmos africanos, a los ritmos modernos, luego a las músicas desestructuradas, después va pasando de las músicas más sensuales a las líricas, para terminar con las minimalistas y llegar al silencio, y en él, a una profunda escucha corporal y emocional.
Después de este viaje, que casi siempre resulta individual, se recoge la escucha, con una palabra o metáfora que nos devuelva al grupo, con la fórmula "me siento como..." y que abre la siguiente serie de trabajos que nos llevará hacia el cuerpo poético.
Por mencionar algunos ejemplos que surgen: *como un volcán en erupción, como un océano, como un acordeón, como una niña en su fiesta de cumpleaños...*

CUERPO POÉTICO

Se trata, llegando a este punto del proceso en que se ha conseguido una cierta flexibilización corporal, de ir un paso más allá, soltando también la mente pensante, para conseguir que el cuerpo se exprese sin los condicionamientos habituales.

Dejarnos sorprender, al igual que perseguimos con otras herramientas artísticas, también por la herramienta artística que es el cuerpo; no es sencillo, porque está habituado a ser regido e incluso dominado por los constructos mentales: para expresar esto tengo que hacer aquello; pero si permitimos que sea el cuerpo el que dirija la acción, es posible que comprobemos que su lenguaje, al igual que el de los sueños, es mucho más rico y sensible de lo que observamos habitualmente.

Se comienza con juegos paradójicos, ejercicios preparatorios de improvisación teatral y de expresión espontánea, todo ello con el fin de entrar en otro registro mental. Repetir una frase desde distintos registros emocionales, contar una historia a-lógica entre todas, relatar una historia común varias personas al tiempo, son algunos de estos juegos. Es interesante observar como a algunas personas "les saca de quicio" este tipo de propuestas, mientras que a otras les apasiona, pero en general, atravesadas las primeras resistencias, resultan divertidos y sobre todo eficaces en su objetivo.

Los preparativos para la fiesta están listos y le llega el turno a la exploración del cuerpo poético.

Igual que un relato puede, por medio de la síntesis, de la extracción de la esencia, convertirse en un mini-relato e incluso en un "nano-cuento", de igual forma, por eliminación de lo superfluo, por condensación, se llega a la expresión de una frase o de una estrofa, en una acción; y de varias acciones surge una narración o un poema.

Requiere un entrenamiento: eliminar los lugares comunes, las expresiones estereotipadas, conectarse con el propio cuerpo y sus muchas posibilidades, estar atento a los sentidos para percibir y poder expresar, y no apremiarse en el hacer, para llegar al movimiento consciente, anclado en un impulso organísmico.

Primero se prueba entre varias personas, con escritos seleccionados por el coordinador o por el propio grupo; más tarde sobre un texto creado por el propio grupo; para terminar cuando el instrumento -corporal, en este caso- está ya afinado, contando mediante una secuencia de acciones, algo personal e íntimo.

En esta ocasión la arteterapeuta seleccionó una poesía surrealista; después el grupo hizo una re-elaboración del cuento de cenicienta, con varios finales para satisfacer diferentes gustos en el grupo; y las acciones de lo personal fueron muy sentidas y contempladas con un profundo respeto por parte del grupo.

Toda esta serie se completa con el ejercicio que da paso al siguiente concepto a trabajar, que es el de Polaridades.

Moveos libremente por la sala, mientras conectáis con la memoria corporal, para que ésta os lleve a través de las sensaciones corporales a aquel momento de toda la experiencia anterior que os llegue como vuestro "momento estelar" aquel en que os habéis sentido plenos. Buscáis en la sala el lugar adecuado y lo realizáis, ya sea una acción, o una sucesión de ellas.

Volvéis a caminar y buscáis el momento de mayor bloqueo, encontráis también un lugar en la sala y lo lleváis también a una acción o sucesión de acciones.

Se trata ahora de ir pasando del movimiento o gesto estelar al de bloqueo, y a la inversa; la clave está en la conciencia, en los micromovimientos que van de uno al otro. Es en el cambio de una postura o de un gesto al siguiente donde está la información.

Este mensaje aportado por lo corporal poético, escuchado con la sensibilidad afinada por todos los ejercicios previos, es incorporado en la memoria corporal, mediante la repetición.

❖ POLARIDADES

El Arteterapia humanista pretende ser una experiencia integradora y para ello necesita superar el mundo de las dualidades excluyentes en que nos educa el pensamiento lineal: bueno-malo, egoísmo-altruismo, femenino-masculino.

Dejar el mundo de lo dual, no significa llegar a ningún tipo de neutralidad que anule las diferencias, sino ser capaz de explorar cada una de las polaridades en nosotros, no como fuerzas antagónicas sino complementarias.

Se trata de reconocer e integrar las múltiples capacidades que tenemos y que nos permiten ajustar nuestras respuestas a las diferentes situaciones que se nos presentan en el contacto con el medio.

Muy a menudo estamos identificados con un extremo de la polaridad, negando o ignorando el otro extremo. Para trabajar con el conflicto interno que produce esta disociación es imprescindible definir bien nuestras polaridades; darse cuenta es siempre el primer paso para salir de esa imagen de nosotros mismos que nos hemos inventado renunciando a otros aspectos que también forman parte de nuestro ser.

Lo que hacemos para trabajar con polaridades, al igual que hace la psicoterapia gestalt, es diferenciar artificialmente —polarizar— para comprender mejor y determinar qué polaridades causan conflicto. Por ejemplo: si pienso como opuestos "racional-emocional" o los conceptos gestálticos de "contacto-retirada" no es que sea mejor una cosa que la otra, ni que uno excluya el otro extremo.

Tan inadecuados pueden resultar el continuo contacto como el mantenerse siempre retirado, tan inapropiado un comportamiento siempre y absolutamente racional como uno emocional en toda ocasión; es la situación la que determina cuando va mejor una actuación u otra. Además entre el blanco y el negro hay multitud de grises; probablemente la respuesta no sea blanco o negro, existen multitud de puntos intermedios además de la alternancia entre uno y otro, y nuestro organismo sabrá escoger el más adecuado a la ocasión, si se lo permitimos.

El trabajo de polaridades lo que pretende es precisamente contrarrestar este empobrecimiento, incorporar esas partes negadas para devolvernos esos recursos que hemos ido perdiendo.

Se trata de que el organismo sea capaz de mirar las dos partes sin juzgarlas, sin negarse ninguna de ellas para poder evaluar y optar. Se trata de saber que podemos ser cálidos y fríos, frágiles y fuertes, adaptativos y creativos, que podemos amar y odiar ...que podemos jugar los dos extremos y todos sus puntos intermedios, porque no los hemos alienado de nosotros.

Un organismo necesita de todas sus capacidades para responder a un ambiente en permanente cambio pero con frecuencia ha ido perdiendo parte de ellas al haberse ido identificando con algunas y rechazando otras.

Y aquí entramos en el llamado *"autoconcepto"*, *"autoimagen"* o selección —interesada— de algunos rasgos de nuestra personalidad con los que nos identificamos, al mismo tiempo que rechazamos otros —otras capacidades— que podrían sernos útiles en determinadas situaciones. Ocurre además que aquello que no reconocemos o aceptamos en nosotros solemos proyectarlo, como ya vimos al trabajar el chakra anterior, sobre los demás.

Se va a trabajar sobre dos conceptos: QUÉ SOY y QUÉ NO SOY; primero se elaboran listas, después se escoge aquella polaridad que esté ocasionando más dificultades en el momento presente.

La dificultad mayor que se presenta al trabajar con polaridades, es polarizar bien, esto es, encontrar los extremos, ya que es habitual que uno de ellos ni siquiera se conciba, por lo que suele ser necesaria la ayuda del grupo y del facilitador. Así ocurre también en esta ocasión con varias personas que cito a modo de ejemplo:

C. Quiere trabajar la polaridad controlador/permisivo, finalmente ve como polaridad descuidador o abandonador/controlador.

M. Ajustando la polaridad tacaña/generosa, la convierte en tacaña/dilapidadora.

E. Buscando su polaridad entre mandona/respetuosa, se encuentra, para su disgusto, con mandona/sumisa

Es importante tener en cuenta en el trabajo con Polaridades que a un extremo no corresponde necesariamente su antónimo, como a "estructurado-desestructurado". Me gusta utilizar para explicarlo la metáfora de los dedos de la mano, tan opuesto al Pulgar es el Meñique como el Anular o el Indice. Cada persona determinará cuál es a su criterio la polaridad; por ejemplo, de amor para unos es odio, para otros es indiferencia y para otros es miedo, porque eso es lo que les impide amar.

Cuando todas las personas del grupo tienen clara la Polaridad que van a trabajar, se les pide que formen tríos, en los que una persona hace de facilitadora, la otra actúa sus polaridades y la tercera hace de testigo con cámara de fotos;

Una vez establecidos los roles de los tríos, actuad la Polaridad de la misma forma en que habéis hecho anteriormente con el "momento estelar" y el de "bloqueo máximo". Llevando cada una de las Polaridades a acción o gesto corporal y explorando también los puntos intermedios entre las Polaridades.

El pasaje de un extremo a otro de la Polaridad, reconstruyendo y deconstruyendo el recorrido, sirve para integrar no sólo los dos extremos sino los pasos intermedios; esto dota a la polaridad de una gran riqueza de contenido para la aplicación posterior en la vida diaria.

Las fotos obtenidas, pasarán a formar parte de su simbología personal, y podrán ser utilizadas si lo desea para el siguiente trabajo sobre la biografía, ya que las imágenes obtenidas en sesiones terapéuticas suelen estar cargadas de significado para la persona.

La propuesta siguiente se relaciona con el tema de las polaridades del amor, y trata de la BIOGRAFIA EMOCIONAL.

❖ BIOGRAFÍA EMOCIONAL

Sobre la importancia de la autobiografía en el proceso de desarrollo personal, quiero decir que casi todas las escuelas centradas en este objetivo, incluyen la elaboración de la autobiografía en alguna de sus etapas. Entre ellas, quiero mencionar como ejemplo de la importancia que se le da a este trabajo a Tara Rokpa therapy, escuela que aúna métodos del Budismo Mahayana con técnicas de psicoterapia occidental; durante el ciclo básico indispensable para pasar a la etapa de entrenamiento como terapeuta, los participantes pasan de uno a dos años escribiendo la autobiografía combinándola con trabajos artísticos y con el objetivo de encarar la inevitable pregunta ¿quién soy yo? ¿Cómo he llegado a ser quien soy? Esas preguntas han sido básicas en la historia de la evolución de la conciencia de la humanidad, y son preguntas sobre las que la escritura de la autobiografía ayuda a reflexionar.

Aunque nosotros no le vamos a dedicar tanto tiempo, sí que ocuparemos un espacio para preparar con la ayuda de herramientas artísticas el clima adecuado para su realización.

FOTOGRAFIA

Ya vimos durante los trabajos con el chakra anterior uno de los posibles usos de la fotografía como herramienta terapéutica, añadiremos ahora otros aspectos.

Es un catalizador que permite descubrir y objetivar aspectos decisivos del proceso terapéutico. Cada fotografía tomada es un "espejo con memoria" que refleja tanto a quien la sacó como ese momento que fue lo suficientemente importante como para ser congelado en el tiempo; estas fotos hacen visible fragmentos de la vida de una persona, resultando ser como huellas que van marcando tanto su recorrido físico como el emocional.

Al observar desde el presente las imágenes del pasado, lo relevante puede estar en:

- Lo obvio que muestra o que no muestran estas imágenes, como gestos, contactos, ausencias repetidas...
- Lo que evocan en la mente y los sentimientos del observador, como actualización de sentimientos que tuvieron lugar en el pasado.
- Los significados que desde el presente el observador proyecta en la imagen.
- Que se hayan convertido en objetos transicionales.

A través de la interacción emocional con estos catalizadores visuales es posible:

- Evocar los recuerdos relevantes.
- Acceder a los sentimientos e información que están contenidas en las imágenes de forma subconsciente.
- Explorar los significados atribuidos desde el presente.

Comienza el trabajo con la Autobiografía emocional con la búsqueda, recogida y selección de fotografías del álbum familiar; esto de por sí ya es un desencadenante de la memoria. A partir de ahí, la elaboración de un collage, con su tarea de pegar, despegar, nombrar, colocar, descolocar y recolocar, permite ir entrando en momentos y relaciones, en presencias y ausencias.

Desde esa atmósfera de creación, se pide que cada autor escriba sobre su vida emocional, narrando los sucesos que en su momento pudieron ser vividos como dramáticos, extraordinarios, vergonzantes, incomprensibles, gozosos, injustos, o cotidianos, pero siempre acompañados de su componente emotiva.

Muchos artistas, como Goya, Van Gogh, y otros, nos han dejado una fiel reseña de su estado físico y emocional en diferentes etapas de su vida a través de sus autorretratos, que son en cierta manera el equivalente pictórico de la autobiografía escrita, por lo que puede ser interesante, que quien así lo desee, incorpore autorretratos, fotografías de obras propias, o de momentos significativos.

Una vez realizada, viene la importante tarea de recoger los frutos que ésta, sin duda, habrá dado.

El efecto de narrar hechos que en su momento pudieron ser dolorosos, situándoles en la globalidad espacio-temporal de toda una vida, y contemplarlos después del tiempo, facilita el que éstos sean vistos desde otra perspectiva, que puede ayudar a dotarlos de significado, relativizar su importancia, contemplarlos con humor, descubrir cómo nuestra biografía ha influido en nuestra personalidad, etc.

Pero además hay que tener en cuenta el potencial que la autobiografía tiene para descubrir los patrones de comportamiento, las repeticiones, el guión de vida que se está manteniendo y que aún, si así se desea, se está a tiempo de cuestionar.

Por ello se aprovecha la autobiografía para hacer un primer abordaje de la localización de los campos en los que con más frecuencia se da el conflicto, intentando reconocer algunas de las pautas de comportamiento que se repiten.

Digo que se hace un primer abordaje, porque en esta ocasión se realiza, como a vista de pájaro, con una mirada global sobre la biografía, mientras que más adelante, dentro de este proceso que seguimos a través de los chakras, se hará de una forma más detenida sobre un período de tiempo limitado, de manera que yendo de lo general a lo concreto nos ayude a afinar el resultado.

La realización de la autobiografía por cuanto supone de esfuerzo, creación y reflexión, sería ya de por sí importante; pero al sumar el aspecto real y simbólico que tiene su entrega a los terapeutas, convierte ésta en un hito dentro del proceso por cuanto tiene también de apertura y confianza. Mientras que para los terapeutas se convierte, además de en una prueba del compromiso, en una ayuda para la mejor comprensión del proceso individual que la persona está viviendo.

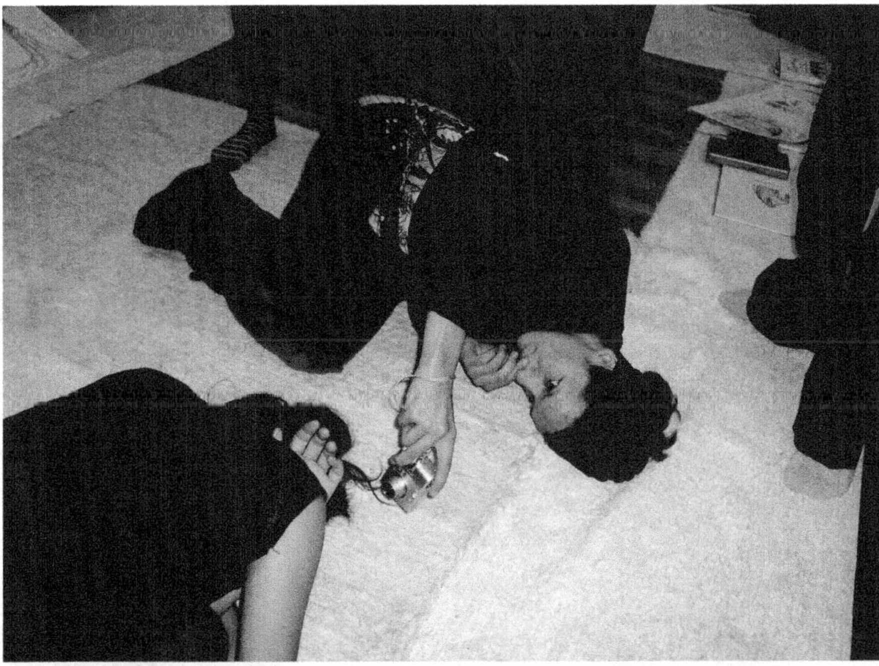

MICRORRELATOS COLOREADOS

VERDE

Como cada primavera, suena el teléfono; oigo la voz nerviosa e imperativa de Falina; después de un rápido saludo me dice "Oye, que las glicinias están a punto de despertar, vente" no hace falta más. Carretera y manta. Y llego a Cadalso, y allí mantenido por el amor de su hija, está la obra de amor por la belleza, de su padre, el palacio de D. Álvaro de Luna que piedra a piedra él restauró. Suerte que era escultor.

Celeste 08

Relativo al Chakra Vishuddha

Escucha: sonidos, músicas, memoria biográfica sonora

Voz

Liberación Cervical Estratigráfica

Ejercicios

— Escucha y Contacto
— Improvisación instrumental
— Expresión por subgrupos: instrumentos + voz + danza
— Bioenergética y expresión emocional
— Retratos sonoros (*feed back* grupal)

❖ CONSIDERACIONES RELATIVAS AL CHAKRA VISHUDDHA

Está localizado en la base del cuello. Incluye la columna cervical, los hombros y los trapecios. Los órganos de la fonación también están regidos por este centro. Es una zona de paso y un centro de control de las energías instintivas y emocionales que ascienden hacia la cabeza, y de las mentales que bajan. El control más conocido es el nudo en la garganta para frenar el llanto o la expresión de la rabia, y los dolores de cuello para soportar lo que no nos gusta. A medida que vamos perdiendo el miedo al descontrol nos vamos enriqueciendo en posibilidades expresivas y de acción.

El color que rige este centro es el azul celeste, y su nombre sánscrito significa "purificación", que podemos entender como el refinamiento y limpieza necesarias (de orden físico, emocional y mental) para facilitar la sintonía vibracional con los niveles más sutiles de percepción. El elemento asociado es el sonido.

Su cualidad es la expresión y la comunicación, por ello se relaciona este chakra con la expresión de la creatividad.

Al llegar a esta parte del proceso trabajaremos desde tres aspectos:

— La escucha: ritmo, armonía, sonido y silencio
— La expresión sonora: instrumental y de voz
— La improvisación, sonora y actitudinal

Psicológicamente tendremos la atención en estas polaridades:

Expresión - Contención
Escucha - Comunicación
Control - Confianza

Son numerosos y lejanos los antecedentes del uso del sonido como herramienta de algún modo terapéutica; está presente en los rituales sagrados, en los arrullos que adormecen a los niños, en la reproducción de los sonidos de la naturaleza, etc.; es decir, está presente en toda nuestra vida, apareciendo como medio de expresión y como vehículo de comunicación.

Los chamanes o curanderos fueron los primeros en usar la música, la danza y el canto como herramientas terapéuticas. Así invocaban a los dioses, se relacionaban con lo trascendente y facilitaban los estados de trance y las curaciones.

Algunos filósofos griegos denominaron a la música "la medicina del alma"; Aristóteles (384-322 a. C.) valoraba la música por su capacidad de facilitar una catarsis emocional.

Durante la Edad Media, el filósofo y matemático zaragozano Ibn Baya (1070-1138) continuó la tradición griega diciendo que la música tiene un carácter que le permite actuar sobre el alma humana, modificando sus sentimientos, sus pasiones y sus afectos.

También Hildegard von Bingen (1098-1179) priora y respetada visionaria alemana, defendía su capacidad curativa.

Lutero (1483-1546) utilizó la música como catarsis, y desarrolló una serie de melodías sencillas para elevar el estado de ánimo de los fieles y con el fin de mitigar el sufrimiento.

Y así, en el Renacimiento, el Barroco, el siglo XVIII, y hasta nuestros días, el interés por el sonido como terapia se ha ido incrementando y afinando hasta llegar al concepto "corporo-sonoro-musical" acuñado por el musicoterapeuta argentino Rolando Benenzon para el cual, el vehículo de comunicación puede ser cualquier tipo de música, grabada o improvisada, pero también los sonidos, y lo que para nosotros es muy importante, sin olvidarse del cuerpo.

❖ CÓMO ESTÁ EL GRUPO

La trayectoria grupal durante los últimos meses ha supuesto una etapa árdua y difícil, en la que han tenido que mirar hacia atrás, abrir el baúl de los recuerdos emocionales y desempolvar sentimientos, para trabajar con la biografía; todas las personas que componen el grupo están satisfechas del esfuerzo que han sido capaces de plasmar en obras plásticas y por escrito, pero hay que decir también que están cansadas, por lo que reciben con gusto la petición de que, para esta ocasión, traigan instrumentos musicales.

Se han trabajado en los pasados meses los contenidos relacionados con los chakras de la zona inferior y media del cuerpo. Les va a tocar el turno a los ubicados en la cabeza, pero antes tenemos que abordar esta región corporal que está en la zona de unión del cuerpo con la cabeza; es un área en la que existen muchos bloqueos, es una de las zonas en la que con mayor frecuencia cortamos nuestra expresión.

Cuando tratamos de evitar la emoción o el llanto, se produce tensión y dolor en la garganta; también cuando tratamos de controlar en demasía, las cervicales se resienten.

Hay otra cuestión que es la escucha; nos cuesta mucho escuchar, al otro, a lo obvio, hemos trabajado la escucha interna, vamos a focalizar ahora la escucha externa, el sonido y el silencio, la escucha y la expresión; ambas se relacionan con esta región corporal.

Otra peculiaridad del trabajo biográfico hecho recientemente, lo constituye el hecho de ser, en gran medida, individual, por ello comenzamos este módulo conectando con la polaridad dar-recibir que ya trabajamos en el módulo anterior relacionándola con la revisión biográfica, planteándola ahora, desde el contenido cuerpo-escucha que nos proponemos abordar. Se trabajará por parejas y grupos, como forma de reencontrar el contacto y apoyo de los compañeros, para hacer el paso del chakra del corazón al de la garganta.

LA ESCUCHA

Hay mucha vestimenta negra en la sala hoy y un aire de pereza, por lo que comenzaremos con ejercicios de activación.

Vamos a ponernos de pie, estirarnos con gusto, dejar salir la voz, con los pies paralelos y poco a poco soltamos los brazos y la cabeza también, botando sobre los talones, dejando que se suelten los hombros, las cervicales y la mandíbula.

Se van proponiendo juegos de relación, en los que con músicas diferentes, se trata de llevar al cuerpo de la pareja, el ritmo del sonido que nos llega; después vienen algunos ejercicios en los que los dos miembros de la pareja interactúan al tiempo, haciendo uno de alga y otro de mar; uno de gotas de lluvia y otro de lago; etc.; se plantean después algunos ejercicios de ritmo, otros de armonía, siempre incorporando escucha y cuerpo, por parejas o tríos; todos son ejercicios de activación, de escucha y de contacto.

Colocamos en el centro de la sala los instrumentos musicales de que disponemos en la escuela, a los que se añaden los que han traído los participantes; es una colección muy sugerente que incluye algunos clásicos, otros étnicos, tradicionales, de música "pobre", y hasta algunos surrealistas. Se les pide que hagan lo que ya muchos estaban haciendo, que es probarlos, e ir escogiendo aquellos con cuyo sonido están más en sintonía en este momento.

Agruparos por cuartetos. Se trata de expresar diferentes emociones que decidáis previamente; hay dos retos en este ejercicio, uno la creatividad, otro escuchar tu instrumento al tiempo que escuchas los demás instrumentos de tu cuarteto.

Esta última consigna no es fácil, requiere un tiempo de ajuste del volumen individual, pero el resultado va llegando poco a poco, es un paralelo de lo que trabajamos desde los primeros encuentros, la escucha interna y externa; y algo más ahora, encontrar el propio sonido dentro del grupo, con todo lo que conlleva de búsqueda de la identidad personal en el encuentro con el otro.

Terminamos esta fase del trabajo de escucha con el encargo de reflexionar por escrito sobre el paralelo de sus comportamientos durante los ejercicios y aquellos que habitualmente tienen en su vida diaria, ¿cómo me he comportado? ¿cómo ha sido mi participación en el grupo? ¿he intervenido? ¿he tratado de imponerme? ¿he escuchado algún instrumento además del mío? ¿he escuchado el conjunto y me he olvidado de mi propio sonido? ¿qué me dice eso de mí? ¿en qué sientes que te beneficia esa actitud? ¿en qué sientes que te perjudica?

LA VOZ

La dificultad del trabajo con la voz radica según Wilfart (1994) reconocido profesor de voz, en que la sexualidad y la afectividad vibran en lugares donde muchas personas tienen problemas no resueltos, y entonces reaccionan rechazando el trabajo, con resistencias diversas, o pretendiendo sublimar sus emociones negativas, en lugar de afrontarlas. Dice Wilfart: *"La riqueza de la comunicación vibratoria reside en un conjunto que partiendo de la sexualidad, alcanza lo espiritual pasando por lo afectivo".*

El origen de los bloqueos en la voz tienen su origen en la ansiedad provocada en el niño, vinculada con el éxito del comportamiento que se le exige; esa ansiedad genera una retención en la respiración. La inspiración se acorta, la energía queda en suspenso en la parte superior del cuerpo, y surge una voz característica de esta abstención; cada golpe emocional que reciba el niño lo registrará el diafragma.

Wilfart nos habla de cuatro diafragmas cuyos ritmos de expansión y de contracción a través de todo el cuerpo están sincronizados; con lo cual el equilibrio entre todos los diafragmas a través de la respiración sería el comienzo de una verticalidad postural correcta, de la posibilidad de vivir las emociones sin que éstas desborden, y en definitiva una armonización de los chakras o centros de energía implicados.

Recuperar el caudal de voz, resulta de una neutralización de lo mental. En la voz hablada, el sonido es continuamente alterado por la ansiedad de la actividad mental y atraído hacia arriba. El ideal consistiría en hablar y cantar con un dominio absoluto, por la parte inferior del cuerpo. Volver a ser capaz, como cuando éramos niños, de hacer vibrar todo nuestro cuerpo para emitir sonidos, esto ocurría antes de que las tensiones acumuladas a lo largo del tiempo, y las represiones emocionales impuestas y aprendidas, impidiesen a nuestro cuerpo vibrar sanamente.

Comenzamos los ejercicios de voz con algunos de precalentamiento que aúnan movimientos corporales con dejar salir la voz pronunciando las vocales; son sencillos y van ampliando la respiración.

> *Continuamos con un ejercicio de bioenergética: pies paralelos, al ritmo de la respiración se mueve la pelvis adelante y atrás acompañándose por el movimiento de brazos y dejando suelto el cuello; en la exhalación se permite salir la voz.*

Este ejercicio es un fuerte activador físico y emocional, por lo que se aprovecha después con un ejercicio introspectivo.

> *Tumbados, cerrad los ojos, sentid cómo vuestro peso cae el suelo, sentid vuestra respiración serenándose... vais pronunciando en voz alta vuestro nombre como lo decían, como os llamaban, distintas personas en diferentes momentos de vuestra vida.*

El trabajo con la voz siempre es difícil, por ello es importante pautarlo cuidadosamente para facilitar su ejecución lo más posible, e ir pasando de soltar la voz, a la expresión de sentimientos.

La energía del grupo ha ido cambiado y ahora está expansiva y cálida. Llega el momento de la improvisación. Pido al grupo que se divida en tres subgrupos: uno va a representar lo corporal que se expresará con el movimiento; otro la parte instrumental, la orquesta; y el tercero será el coro.

La orquesta escoge los instrumentos que van a utilizar y se colocan en un extremo de la sala; en el otro extremo los que se expresarán desde el cuerpo, la coreografía, de pie repartidos por el espacio; y entre unos y otros, bien agrupados, el coro.

> *Todos en silencio, os pido mucha escucha interna y externa. Cuando alguien sienta que quiere comenzar, inicia ya sea un sonido, ya un movimiento, y quien desee acompañarle, bien sea con un movimiento o un sonido, lo hace; y así permitid que se vaya tejiendo un trabajo colectivo.*
>
> *Recordad que no hay mal ni bien, lo que surja es lo que tenía que surgir. El grupo es más que la suma de sus partes, no estáis contando historias separadas sino una común.*

Se repite intercambiando los papeles para que todos tengan la oportunidad de experimentar en los diferentes roles. Cada vez resulta diferente y les sorprende por su armonía, pero ese es el efecto de todos los trabajos previos realizados más la concentración que la propia improvisación requiere.

Continuando con la improvisación, ahora que el grupo se siente más unido y reasegurado en su capacidad, llega el momento de pedir un mayor compromiso. Les pido que se organicen en cuartetos y que se repartan por la sala.

RETRATO SONORO

En cada cuarteto alguien se ofrece para ser retratado, las otras tres personas debaten en su presencia como qué tipo de música le perciben, con cuáles instrumentos, sonidos o voces, con qué ritmo, con qué intensidad, etc., y se lo ofrecen.

Por turnos todos van pasando por el rol de retratado. Es emocionante estar presente cuando la creatividad se manifiesta.

Estos son algunos de los comentarios recogidos:

B: Me han hecho como una tormenta en plena naturaleza y luego me han acompañado a la calma, a la playa...

M: Me han regalado una pieza de fuerza, de intensidad, de alegría que sin perder el ritmo, sin perder el swing, va avanzando hacia algo más sutil.

S: Me han clavado, tengo la sensación de que me han clavado. Primero un pataleo, con maracas, que va bajando de intensidad hacia un sonido suave de garganta como un ronroneo.

C: A mí me han retratado muy espiritual y ordenado y me han ido desordenando, me han llevado desde Oriente a África.

E: Han hecho una composición misteriosa, con suspense y contrastes.

L: Me han retratado como un ritmo suave, que arropa de una forma maternal y contenida, y me han regalado un final con una voz potente y grito incluído. Justo lo que me hace falta.

Terminamos esta fase del trabajo con la voz, con una reflexión escrita y compartida que tiene como finalidad, entre otras, que se abra al grupo cualquier agradecimiento o posible malestar no expresado ¿cómo han sido los cambios de mi voz? ¿cómo me siento con las devoluciones que me han hecho?

❖ LIBERACIÓN CERVICAL ESTRATIGRÁFICA

En la vida todo es ritmo, desde el latido cardíaco a los ciclos del sueño o de la naturaleza. Cuando estamos en sincronía con el ritmo de otra persona, de una música, o de cualquier experiencia, tendemos a permanecer "en fase", decimos que resonamos, tenemos la impresión de comprenderla profundamente, de sentirla. Se acostumbra a poner como ejemplo, el hecho de que si tenemos dos violines de igual afinación podemos hacer que uno de ellos vibre, tocando una nota en el otro violín. Este hecho tomado como metáfora del trabajo terapéutico, implica que podemos activar un estado de atención y de conciencia en otra persona, utilizándonos a nosotros mismos como instrumento.

La experiencia que se propone a continuación, requería de toda la preparación anterior e incluso de los meses de trabajos grupales que han ido creando la confianza personal y en el grupo. Aúna ejercicios de desbloqueo cervical, diferentes tipos de respiración, capacidad para el maternaje y sobre todo confianza.

La denominamos "estratigráfica" porque mediante los sencillos movimientos que se van haciendo y el apoyo de la respiración, se van atravesando diferentes estratos psíquicos; cada persona a su ritmo y con sus condicionantes, va deshaciendo bloqueos y, con suerte, entrando en contacto con nuevos estratos.

> *Poneros por parejas, uno va a hacer de facilitador y el otro de cliente; el cliente va a recibir un masaje relajante por parte de su pareja.*
>
> *Después el facilitador se sienta junto a la cabeza del cliente, para ir realizando con mucha atención, con actitud de cuidado y disponibilidad, los movimientos cervicales que se le irán indicando.*
>
> *En algún momento del proceso aparecerá un cambio en la respiración del cliente que habrá que apoyar, y a partir de ahí los terapeutas que van a estar todo el tiempo por la sala, os ayudarán.*

Hay un clima expectante, aunque se les dice a los clientes que no esperen nada, que simplemente se entreguen a la experiencia; a los facilitadores se les pide que solamente estén con presencia, atentos a sí mismos y a su cliente. Surgen, como en cualquier situación inhabitual, las inseguridades personales, pero también una disposición muy importante de entrega a la experiencia.

No voy a describir el proceso que es lento y único para cada persona, pero sí decir, que tan importante y desvelador resulta para el cliente como para el facilitador; que es difícilmente racionalizable, aunque, a la postre, todos lo intentamos; y que ayuda a que emerjan, según cada caso, desde emociones que parece que viniesen de etapas muy tempranas, a estados extáticos o las llamadas experiencias cumbre, que tienen el valor de darnos una nueva perspectiva sobre nuestros conflictos presentes. Insistimos siempre en que no hay vivencias más valiosas que otras y que cualquier cosa que se haya vivido es aquello que se necesitaba en ese momento.

JAM SESSION

Para terminar los trabajos con este chakra, realizamos una última sesión de improvisación para que cada persona tenga la oportunidad de explorar su nueva identidad sonora, a través de lo corporal, vocal e instrumental; y también, y esto es muy importante a esta altura del proceso de grupo, para que tengan la oportunidad de afianzar o encontrar un nuevo espacio en el grupo: más discreto, más protagonista, más suave, más confrontativo, etc.

Se comienza probando la identidad sonora de cada persona;

> *Cada persona se expresa con el instrumento, sonido e intensidad que desee.*
> *Ahora, preguntaros si queréis añadir algo a vuestra expresión habitual: sutileza, fuerza, alegría, atrevimiento, juego, seriedad...*

Después de escuchar a cada persona con su instrumento, van eligiendo quien siente que tiene que empezar, quien quiere seguir...

> *Comienza uno a tocar, se incorpora un segundo, después un tercero, y cuando ya estéis todos incorporados comienza el juego; quien quiera toma protagonismo, quien quiera acompaña, quien lo desee sube de intensidad, o baja, o cambia el ritmo, o lo mantiene, completamente libre. Disfrutadlo.*

Siguiendo ese orden, se van sumando los instrumentos. Consiguen que se escuchen todos, porque ellos mismos se están escuchando a sí y a los otros.

Le sigue un largo silencio, al final del cual pido que expresen con una palabra cómo se sienten en ese espacio nuevo personal y en el grupo.

Esta es parte de la cosecha:

> *Deshielo, ilusión, sosiego, calma, apertura, fuerza, identidad, integración, continuidad, seguridad, confianza, unión.*

Siento fuertemente la cohesión grupal en este momento, y también un sentido de trascendencia.

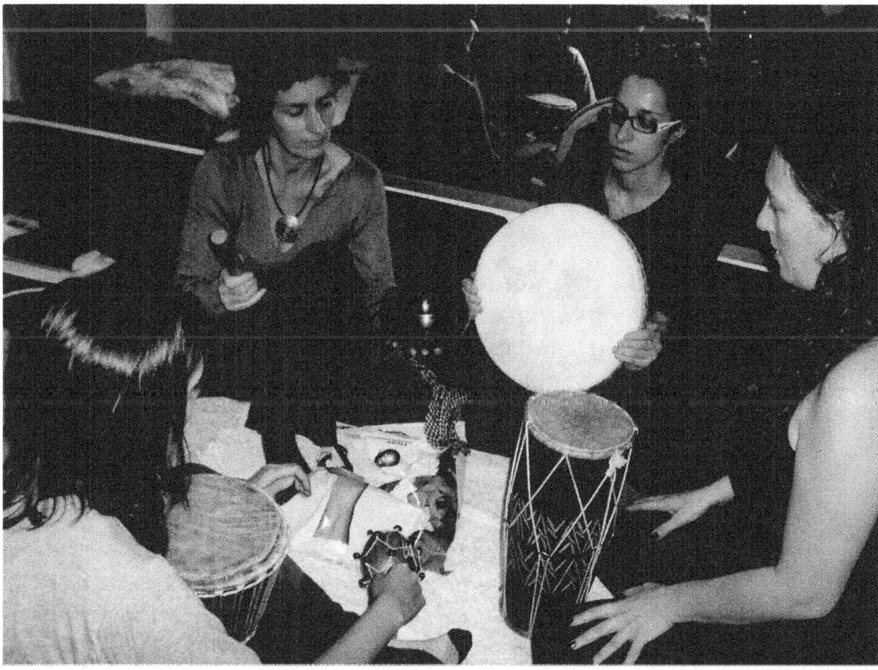

MICRORRELATOS COLOREADOS

CELESTE

Estaban en la galería de arte, era la inauguración de la exposición de sus últimas obras.

La autora hablaba de que la creación para ella era como detenerse fuera del tiempo, hasta escuchar algo que la memoria parecía querer recordar, que aparecía entre niebla hasta que por un leve instante un rayo de sol lo iluminaba. Es efímero y hay que cazarlo al vuelo, dijo.

El crítico de arte a su lado respondió: "Pero los resultados a veces son como gritos de rebeldía, recuerda los muralistas mejicanos, o gritos de ira, como en el Guernica, y hay otros que por muy grandes que sean o están mudos o hablan con la voz estrangulada, y bueno hay algunos que, reconozcámoslo, no dicen más que tontunas".

Relativo al Chakra Ajna

Armonización de funcionamientos

— Lógico / analógico
— Racional / emocional

Análisis del Proceso Personal

Reconocimiento de Pautas de comportamiento

Ejercicios

— Mandalas
— Marionetas
— Dramatización de sueños

❖ CONSIDERACIONES RELATIVAS AL CHAKRA AJNA

Lo primero que hay que tener en cuenta es que en lo que respecta a los dos últimos chakras, sexto y séptimo, existen algunas disparidades de criterio según las fuentes que se consultan en lo relativo a su ubicación exacta y al color con que se relacionan; no obstante no existen diferencias significativas en cuanto a sus funciones.

AJNA, significa "percibir" y "ordenar".

La ubicación del sexto chakra, de forma simplificada, se dice que es, unos centímetros por encima del entrecejo y hacia el interior. De forma mas precisa, para localizar este centro, deberíamos trazar una sección por el interior de la cabeza que uniese la frente por encima de los arcos ciliares con la zona del hueso occipital (esto ocurre con todos los chakras, que afectan por igual a la parte anterior y posterior del cuerpo, por ello algunas teorías apuntan hacia una duplicación de centros).

Sus funciones: ver (física y psíquicamente), capacidad mental de abstracción, de lucidez y de orden. Equilibrio. La percepción de colores e imágenes internas, entre las que se encuentran las producidas durante el sueño.

Está relacionado con nuestra percepción del tiempo: las imágenes de los recuerdos, y las del porvenir. Desde esta facultad, se pueden reconocer con mayor facilidad las pautas, identificar comportamientos o situaciones aparentemente nuevas que entran en resonancia con otras anteriores.

Su desequilibrio afecta físicamente a: problemas oculares, jaquecas, pensamiento confuso, pesadillas, alucinaciones, las dificultades relacionadas con el sistema nervioso.

El color con el que se asocia con más frecuencia es el Indigo, un azul oscuro con un tinte de carmín o púrpura; es tan oscuro que se aproxima a un negro con matices, poéticamente se describe como el cielo en una noche de verano.

Tiene que ver con la armonización de los dos hemisferios cerebrales, o mas ajustadamente, de las funciones asociadas a cada uno de los hemisferios:

IZQUIERDO – YANG	DERECHO – YIN
Lógica	Sentimientos
Detalles	Visión de conjunto
Lenguaje, palabras	Símbolos e Imágenes
Presente y pasado	Presente y Futuro
Matemáticas, Ciencia	Filosofía, Arte
Comprender	Percibir
Información	Experiencia
Percepción de modelos y relaciones	Percepción espacial
Nombre de los objetos	Función de los objetos
Práctico	Impetuoso
Racional	Emocional

❖ CÓMO ESTÁ EL GRUPO

Cuando llego a la sala noto un ambiente más silencioso de lo habitual; ante la propuesta de que digan cómo están, nadie parece tener muchas ganas de hablar. Espero mientras observo sus posturas y expresiones; las miradas perdidas o dentro de sus cuadernos me indican que aún no están preparados para compartir su mes, ni su momento; voy recordando al mirar a cada persona del grupo sus vivencias en el último encuentro, la mayoría fueron emocionalmente intensas, tengo ganas de saber cómo lo han digerido durante el mes, qué han hecho con aquello tan significativo para ellos en aquel momento.

La primera en hablar es R. Mientras empieza titubeante contándonos como le fue las primeras semanas después del encuentro, me vienen al recuerdo imágenes de un trabajo que hizo hace unos meses; era una dramatización en relación con la figura de su padre, un hombre al parecer con una personalidad brillante y con vocación de ser el centro de cualquier reunión, a quien ella admira y rechaza a partes iguales; se había visto a sí misma observando desde lejos alguna escena infantil en la que su padre ejercía el protagonismo en un grupo mientras ella, desde la invisibilidad, le contemplaba sintiéndose ignorada y celosa. Hasta aquí era lo que ella recordaba en ese momento; continuamos el juego cambiando los roles, invitándola a que ocupase el centro de la escena, y después de algunos remilgos pudorosos, allí estaba ella, poderosa, disfrutando de un rol al que en su momento no tuvo acceso, y del que no había vuelto a hablar; ahora después del trabajo con la voz, dice:

> *R.: Ha sido muy desconcertante para mí que las primeras semanas después del taller, me sentía super-cómoda haciendo de protagonista con los compañeros en el trabajo, con los amigos; bueno, no siempre me atrevía, pero sí me doy cuenta de que me apetece y me gusta; lo que me deja un poco descolocada es que yo siempre había criticado a las personas que lo hacían.*

Es importante recoger que esto es algo que ocurre con frecuencia, parece que un hallazgo pasa al olvido pero sigue "cociéndose a fuego lento", continúa haciendo su proceso, y un tiempo después reaparece evolucionado.

Siguen los demás compartiendo con el grupo sus vivencias; en general ha sido un mes con altibajos emocionales, eso explica la pereza de hablar, les cuesta identificar lo vivido, aparece varias veces la frase "ha sido un mes muy movido", pienso que les va a venir muy bien el trabajo fundamentalmente cognitivo que van a hacer al comienzo de este módulo.

Les explico que el Arteterapia desde la perspectiva del chakra Ajna significa desarrollar e integrar las capacidades y funciones de los dos hemisferios cerebrales: lógica, pensamiento, acción, de un lado, y pensamiento analógico, emoción y receptividad, por otro lado; y que vamos a aplicarlo a su proceso personal, ya que, después de todo el trabajo de autoconocimiento que han realizado a lo largo de este curso, es el momento de evaluar, sacar conclusiones y elaborar un mapa de su recorrido. Para ello, me revisto de un halo lo más analítico y didáctico de lo que soy capaz y dibujo verbal y visualmente esquemas aclaratorios sobre el tema.

EL PROCESO PERSONAL estudiado aplicando nuestra capacidad de análisis racional implica:

- Observación de datos.
- Ubicación cronológica de los mismos.
- Consideración de las ideas-valores-prejuicios que han podido motivar determinadas acciones.
- Apreciación objetiva de la situación actual.

EL PROCESO PERSONAL visto a través de nuestra capacidad analógica y emocional implica:

- Observación de sincronicidades, relación de situaciones vitales con acontecimientos aparentemente fortuitos.
- Repeticiones de hechos, o de sucesión de acontecimientos similares.
- Emociones que se presentan con mayor fuerza y frecuencia, y ante qué situaciones o tipologías personales aparecen.

TODO ELLO DARÁ COMO RESULTADO:

- Reconocer el momento personal en que nos encontramos durante el proceso terapéutico.
- Observar la clase de contenido que emerge de nuestro fondo psicológico.
- Identificar las emociones que afloran con más frecuencia en los intercambios relacionales.
- Saber las acciones que están más acordes con las habilidades previas.
- Saber también, los asuntos inconclusos, ya reconocidos durante el proceso, así como las repeticiones.

El trabajo que se va a abordar, es una especie de balance entre la persona que comenzó el proceso y el yo renovado que aparece en el presente; hacerlo implica ser capaz de una mirada desidentificada, una visión como observador o testigo del propio proceso.

Después de repartir cartulinas, rotuladores y ceras se les pide:

> *Dividid el papel en tramos que representen los meses desde que empezó el curso, e id anotando las situaciones más significativas vividas, representando con colores el tipo de implicación mental, corporal o emocional que acompañaba a la situación y cuál era vuestra sensación o emoción;*

Una vez elaborado el mapa del recorrido vital de estos últimos meses, vamos a analizarlo de una manera distanciada.

❖ RECONOCIMIENTO DE PAUTAS

- Localización de campos en los que repetidamente se nos da el conflicto: pareja, trabajo, salud, amigos...
- Encontrar la secuencia de repetición.
- Descubrir el arquetipo que hay detrás y cuándo surge.

EL GRUPO. Algunas cosas que ocurren.

El peligro de esta parte del trabajo es que con frecuencia aparece la tendencia a generalizar, y desde ahí no hay posibilidad después, de conectar con las emociones que acompañaban a los sucesos; por lo que es importante insistir una y otra vez, en encontrar situaciones concretas que lleven vida en su interior, para poder descubrir desde qué lugares emocionales son vividos los acontecimientos. ¿Qué sentimientos son los que más se nos repiten? ¿en qué tipo de situación? ¿con qué personas?

I. dice que hay algún período del que no guarda memoria de nada significativo; le recuerdo que cuando escribió la biografía emocional que pedimos hace unos meses, le ocurrió algo similar y la pregunta obligada es ¿qué te puede decir eso de tí como pauta de comportamiento?

C. ve su secuencia de repetición en la relación con su jefe, no llega a decirle lo que quiere y se queda en la rabia; en ocasiones le pasa lo mismo con su suegra; su fantasía, es decir, la imagen del yo que hay por debajo, es que no va a saber hacerlo sin ponerse muy agresivo.

M. observando el esquema resumen de sus vivencias en estos últimos meses se da cuenta de sus mandatos intrapsíquicos "Para que me quieras tengo que cumplir tus expectativas, antes incluso de que tú me pidas nada, tengo que adivinar tus deseos", dice, "este comportamiento lo repito una y otra vez con personas y en ámbitos diferentes, esto no sólo me tiene harta sino que después hace que me quede rencorosa". Debajo de esta pauta repetida de comportamiento, subyace una imagen del yo ¿cuál crees tu que puede ser? le pregunto; A lo que sin dudarlo, responde "Que no me creo digna de que me quieran por mi misma, me lo tengo que ganar".

Algunos van identificando ya, la pauta de comportamiento, con su secuencia y también la imagen o arquetipo que está detrás y que es la causante del comportamiento, pero no todos.

Para completar el ejercicio y facilitárselo a las personas que son más visuales propongo en otra cartulina elaborar un Mandala: un dibujo en el que se representa en el centro a la persona que lo está realizando y a su alrededor, las diferentes áreas de su vida: familia, trabajo, amigos, el grupo de arteterapia,..

Dibujad primero un círculo en la cartulina, en el centro un dibujo o forma que te represente, y alrededor, la familia, trabajo, amigos,.. simbolizando mediante el tamaño, y la forma, el espacio que te han ocupado en los últimos meses, y mediante el color las emociones con las que lo has vivido. Mientras que lo realizáis, procurad sintonizaros con situaciones concretas, para conectaros con la emoción y recordad también cuál fue la secuencia de repetición.

Con este ejercicio han pasado de una representación temporal a una imagen globalizadora; para algunos ha sido más fácil y para todos ha completado el ejercicio anterior.

C. vuelve a intervenir ahora concretando más, en la situación que exponía antes con su suegra, le gustaría poder decirle que no se metiese en su vida, pero además de temer no saber decírselo de una manera adecuada, teme las consecuencias de después, entonces se calla y se queda molesto, haciéndoselo sentir de alguna manera indirecta.

I. ahora ya lo ve con más claridad; cuando la relación con una posible pareja empieza a plantear dificultades, deja de resultarle interesante, se enfría emocionalmente, y pasa a otra cosa. "Para evitar el esfuerzo de tratar de resolver el conflicto o quizás fantaseando que no va a ser posible, abandono, y me olvido del asunto" y añade "es posible que lo haga también en otras circunstancias pero como lo veo más claro es con las parejas".

Ahora que ya tienen claros los personajes internos, es el momento de ponerlos fuera, encender las candilejas y verlos actuar.

Han jugado mucho con nosotros y ahora vamos a jugar nosotros con ellos.

❖ TÍTERES

Iniciaremos la elaboración de los títeres para después preparar una actuación con ellos. El trabajo de arteterapia con títeres, al igual que el que se realiza con teatro, tiene la particularidad de la observación de las relaciones que se establecen entre diferentes personajes que representan mecanismos psicológicos que la persona utiliza cotidianamente. Particularmente en el espacio simbólico de los títeres, parece que es más fácil observarlos distanciadamente ya que hay menor implicación emocional que cuando se dramatiza y uno "se mete en el papel", y es desde esta desidentificación del observador que se facilita la observación y comprensión del juego en el que estamos metidos.

ALGUNOS TIPOS DE TÍTERES

Títere plano: Puede ser desde el más sencillo de elaborar que se adapta bien a niños pequeños y se hace con cartulina recortada, al mas complejo para adultos o jóvenes, que se puede realizar con madera o cartón grueso, recubierto de materiales diversos, y articulado, como los javaneses.

Títere de guante (guiñol): La cabeza se mueve con el dedo índice y los brazos con el pulgar y el central; se puede recrear-transformar a partir de una ya prefabricada o bien, si se desea hacerla, los soportes para elaborar la cabeza que deberá ser hueca, pueden ser, un globo, una botella de plástico, un modelo en arcilla..., que después se recubre de papel-maché. Después atender a las manos y a la vestimenta y accesorios.

Títere de hilos o marioneta: Presenta mas dificultades de elaboración y de manipulación, puede ser muy interesante para trabajos en grupos.

Títere habitable: Por una o varias personas que se introducen dentro de él; es un objeto a animar. Como ventaja cuenta con la posibilidad de diferentes tamaños, como inconveniente la despersonalización.

PERSONAJE Y VOZ. *La propuesta es que partáis de los personajes que os habéis descubierto en el trabajo anterior, que les pongáis un nombre, y que les empecéis a dotar de características propias; aprovechando que habéis hecho ya los módulos de voz, dotadles también de expresión a través de darles voces diferentes.*

MONÓLOGO. *Se pide comenzar por un monólogo de presentación de cada personaje, lo que aporta información sobre motivaciones, justificaciones y ansiedades de cada uno de ellos.*

EL GUIÓN. *Una vez que tenéis los personajes, el siguiente paso es crear la historia, pensad en un argumento que puede ser simbólico o apoyaros en un conflicto con el que lidiáis en la actualidad, y mostrad a través de los diálogos del guión cómo actúan vuestros personajes.*

LA REPRESENTACIÓN. *Ante un público entusiasta compuesto por sus compañeros y terapeutas, se lleva a cabo la representación.*

El público conforma una audiencia activa, ya que al finalizar puede preguntar a los personajes sobre sus motivaciones, emociones...

Durante la representación aparecen muchos factores dignos de observación, como:

— el tipo de títere y técnicas elegidos,
— dificultades y capacidad de improvisar durante la ejecución,
— la elección de las características del personaje en relación con la historia en su conjunto,
— la adecuación y versatilidad de las voces empleadas, y
— las resistencias.

Vale la pena comentar algunas de las resistencias aparecidas, ya que a esta altura del proceso después de haber visto y haber hablado de sus mecanismos de defensa, sus polaridades y tantas otras cosas, todas las personas que componen el grupo son capaces de distinguir sus personajes internos; sin embargo, darles cuerpo y voz, ponerles tan abiertamente a la intemperie, les provoca múltiples sentimientos; no es un trabajo que se pudiese hacer antes.

En cuanto a proceso personal, requiere haber ido descubriendo paso a paso diferentes aspectos personales y haber ido tomando confianza con el grupo hablando de ellos, y escuchando cómo los demás hablan de los suyos, para llegado este momento poder prestarse al juego. Aún así, después de varios meses de trabajar con ellos, en el último momento, justo en la representación, aparecen las resistencias:

D. es una persona muy cumplidora en cuanto a asistencia y entrega de trabajos, por ello nos sorprende especialmente que envía sus marionetas de la mano de M.; son tres, miden unos 40 cms., están magníficamente elaboradas, se ve el tiempo que les ha dedicado en los detalles y características diferenciadoras de cada una de ellas, pero una gastroenteritis de última hora le ha impedido, o le ha facilitado, no presentarnos ella el guión.

J. presenta unos títeres que podríamos denominar minimalistas, los ha construído de forma cuidadosa con objetos de reciclaje: un cartón de leche, una ratonera, tornillos, etc.

J., comenzó en el grupo con dificultad para hablar de sí, y después de tantos meses y tanto afecto manifestado por sus compañeros ha conseguido la confianza necesaria para abrirse; en esta ocasión trae una historia que habla de sí y de sus personajes, pero en un tono tan simbólico que la mitad del grupo no consigue entender.

L. a la mitad de la representación olvida el guión según dice después, y asistimos a la representación de un acting-out, su historia se convierte en una pelea de estacazos y voces, que recuerda los títeres infantiles, cuando niños y marionetas se alían para matar al lobo malo, y que consigue dejarla fuera del contacto emocional.

También ocurren imprevistos que se podrían llamar bendiciones:

C. uno de sus títeres durante la representación, sufre un pequeño accidente y pierde la cabeza; los espectadores se ríen por el paralelismo del fuerte peso que tiene lo mental para C. y él comienza a improvisar como si realmente hubiese perdido si no la cabeza al menos el guión; cuando termina se le ve excitado, divertido, y asombrado por los diálogos que han surgido de sus personajes.

Después de que Winnicott ampliase la visión de la psicoterapia definiéndola como "*dos personas que juegan juntas*", se han utilizado con mayor frecuencia los títeres también con pacientes psiquiátricos adultos, en particular con psicóticos, apoyándose en la idea de que éstos, los psicóticos, tienen una facilidad particular para proyectarse en el espacio que les rodea, de forma que fabricar un objeto es una especie de recreación de sí mismos. Hasta ahí, es lo mismo que para cualquier producto artístico.

Pero los títeres además de permitir ser objeto de identificación proyectiva y de tomar el valor de *objeto transicional* (Winnicott), facilitan la relación entre diferentes partes del sí-mismo y la relación con el otro.

Tanto en el caso de dificultades de relación grupal, como cuando la relación cliente-terapeuta está bloqueada, el títere puede funcionar como un objeto intermediario que permite establecer una comunicación, como cuando el cliente desde su marioneta se dirige al arteterapeuta o a alguien del público, o cuando el arteterapeuta se dirige a alguna de las marionetas sabiendo que lo está haciendo a los dobles parciales del cliente.

Si fuésemos capaces de ver el mundo que conocemos como un gran teatro, donde cada uno vive su arquetipo favorito y puede intercambiarlo por otro cuando el guión así lo requiera, podríamos tener una visión muy nueva y distinta de la vida.

Los personajes están interrelacionados entre sí: un sheriff no tiene razón de ser si no hay bandidos. Luego, los personajes se atraen entre sí. El personaje con el que te identificas fatalmente se encontrará con otro que le dé sentido. Lo que podemos hacer es escoger un personaje que nos haga agradable el teatro. Pero sabiendo que no somos él, y que para cambiarlo, basta con cambiar de máscara y de traje. En otras palabras ¡dejar de identificarnos con los personajes!

Trabajar desde la perspectiva de este chakra me suele producir una sensación de "completud" que es una palabra bastante poco usual, supongo que no solo lingüisticamente, sino también por lo que significa. Analizar por un lado nuestros procesos para descubrir las pautas de comportamiento da una cierta sensación de poder, de "ahora que lo he entendido podré manejarlo mejor". Por otro lado ser capaz de hacer caricatura a través de las marionetas de nuestros personajes internos, es podernos distanciar de ellos a través del humor, y eso nos expande, nos hace ver que somos más que aquello de lo que nos reímos. El trabajo con los sueños es la parte que completa la trilogía, es la oferta de información nueva que nos llega a través de un registro que no podemos controlar, y desde ahí es un espacio limítrofe con la magia, por cuanto tiene de desconocido u oculto.

❖ EL TRABAJO CON SUEÑOS

Los sueños son vistos como proyecciones de la personalidad del soñante, que se manifiesta en las imágenes oníricas con uno o varios significados y con mensajes existenciales. Jung decía que casi todos los sueños se ofrecen a ayudar al soñador en la tarea de vivir.

Una teoría explica el proceso de los sueños de la siguiente manera: Durante las primeras horas del sueño la mente se dedica a procesar los elementos emocionales que se han activado durante el día y que no se han podido elaborar adecuadamente. Esta parte del sueño suele estar compuesta por personajes y escenarios conocidos; son los sueños compensatorios. Después aparece la actividad onírica propiamente dicha, o los sueños no compensatorios, con imágenes que provienen del inconsciente.

Otra teoría explica el proceso como de tres horas de duración con tres puertas que se abren, una cada hora, dando acceso a diferentes partes de la actividad: la primera en la que se presenta el sueño con imágenes fijas, la segunda, en que se plantea el conflicto, y que en ocasiones genera sobresaltos, y la tercera en la que aparece la resolución del sueño.

En mi experiencia personal he podido observar que en algunos trabajos con sueños, se aprecia que la persona recuerda varios episodios oníricos en la noche, y al reconstruirlos se comprende que plantean el mismo conflicto con metáforas diferentes.

Los sueños hablan de aspectos que en muchos casos, son desconocidos para el "yo consciente" por lo que constituyen un importante recurso clínico y cuando se ignoran, en ocasiones se convierten en pesadillas, como único recurso del subconsciente para hacerse oír.

Los "sueños incubados" que responden a una pregunta formulada la noche antes, tienden a ser más fáciles de recordar y más vívidos que los sueños espontáneos.

Con el trabajo de sueños en Arteterapia Humanista se persiguen dos objetivos:

1. Facilitar que el cliente determine cuál es el mensaje existencial que conlleva su sueño.
2. Reincorporar dicha experiencia enajenada a su personalidad.

3. Desvelar el imaginario personal, asociaciones, símbolos y metáforas que afloran del inconsciente.

El precursor del trabajo con los sueños mediante la Dramaterapia fue Jacob Levi Moreno(1923), poeta, médico, conocedor de la filosofía de las religiones, psicoterapeuta, y hombre de teatro influenciado por Stanislavski.

METODOLOGÍA

– En el trabajo de elaboración de los sueños a través del Arteterapia, éstos, al igual que toda experiencia, deben ser experimentados en lugar de explicados.

– Es importante resaltar que sólo el productor del sueño puede interpretarlo, sólo quien lo crea tiene las claves de su significado. La misión del arteterapeuta, es la de ayudar a hacer las preguntas necesarias, crear una atmósfera de aceptación, interés y aventura compartida. Además de esto, debe de estar lo bastante seguro de sí, como para admitir que el soñador, y nadie más que él, puede suministrar la información necesaria para comprender el significado específico del sueño. Digo esto porque cuanta más inseguridad tiene el terapeuta más suele empeñarse en demostrar que "sabe", que dispone de claves que el soñante no tiene, cuando su labor es justamente dar confianza al soñante para que vaya descubriendo las pistas que su subconsciente le ha ido dejando.

– Admitiendo que no hay coincidencias en los sueños sino que por el contrario hemos elaborado el guión prestando atención a cada detalle, tendremos que todo elemento del sueño puede ofrecernos claves sobre su significado. Cada elemento es visto por tanto como expresión del soñante susceptible de ser trabajado como proyección mediante la técnica de "la silla vacía", esto es, por identificación del soñante con cada elemento, diálogo entre las partes y reapropiación del contenido.

– También se elaboran los sueños en que aparecen personajes, dramatizando la escena con la ayuda de los participantes en el grupo.

– Ayudarse del dibujo o la pintura suele ser muy útil, cuando aparece algún elemento rico en simbología sobre el que entendemos que vale la pena profundizar.

- Utilizamos técnicas propias de la *Instalación* y *Performance* cuando por la distribución de los personajes o elementos en el espacio, éste parece ser una clave importante en el significado.

- Cuando en el sueño se presenta un dilema, que puede reflejar el dilema vital que vive el soñante, exploramos mediante la dramatización posibles continuaciones del sueño.

- Cuando lo que aparece en el sueño es uno o varios arquetipos se profundiza en el significado personal de éstos.

- Siempre se comienza por la narración en presente y primera persona del sueño, y se pide que el narrador esté atento a las emociones que le van surgiendo y las comunique.

Resumiendo, en el trabajo con sueños desde el Arteterapia Humanista, empleamos técnicas propias de la Gestalt, del Psicodrama de Moreno, y cualquier expresión artística que entendamos, en cada caso, que puede facilitar la comprensión del mensaje existencial.

Los discernimientos, ideas, o soluciones que se reciben después de un sueño tanto incubado como espontáneo, o tras una meditación, pueden llegar en forma de corazonadas o impulsos que, por supuesto, deberán ser revisados empleando nuestra racionalidad y sentido común antes de ser llevados a la práctica.

CÓMO ESTÁ EL GRUPO

El trabajo con sueños comienza cuando se les pide que los que tengan sueños recurrentes tomen nota de ellos y en cualquier caso, que todos estén atentos a los sueños que recuerdan durante el mes y que los escriban. Por lo general, el hecho de poner la atención en ello es suficiente para que la mayoría de personas sean capaces de recordar al menos fragmentos, tan útiles para el trabajo como los sueños completos.

Después de los ejercicios más cognitivos sobre las pautas de comportamiento, esta propuesta adelantada de fijar la atención en lo onírico es recibida por el grupo, como la primavera después del invierno.

Sueño n.º 1 - EL MOCHILÓN

Estoy subiendo a un barco por una pasarela estrecha, en escalera, llevo una mochila que transporto cómodamente en mi espalda, hacia la mitad de la escalerilla miro para atrás y veo a mi pareja que, cargada con un gran mochilón en los brazos que le impide ver los escalones, no consigue subir. Retrocedo algunos escalones, me quito mi pequeña mochila de la espalda poniéndomela delante, y le pido que me dé la suya que cargo a mi espalda. Continúo subiendo, ahora con gran dificultad, no solo por el peso sino porque la mochila grande que llevo a mi espalda va chocando contra las barandillas. Cuando por fin llego al barco y me descargo del mochilón ajeno, miro a mi pareja que está terminando de subir la escalera y me despierto con una enorme sensación de enfado.

Este sueño parece que tiene que ver con el trabajo de la sesión anterior, con un tema de repetición de pautas de comportamiento, y así lo identifica la soñante que a través de la emoción con la que se despertó pudo darse cuenta de cuánto de enfadada estaba con su pareja por lo que ella entendía como que estaba cargando con responsabilidades que no eran suyas. Diremos que su pareja, separado, tenía dos hijos de los que ella se estaba ocupando, al parecer más de lo que quería. La metáfora de la mochila era tan gráfica para la protagonista al representar las cargas que no pareció necesario trabajar sobre ello en ese momento. Por lo que centramos la elaboración en ver qué quería hacer con lo que había experimentado tan claramente en el sueño. Su enfado con su pareja.

A este tipo de sueños, les apodo familiarmente "sms del subconsciente" porque son tan claros para el soñante que apenas requieren elaboración, su función es decir alto y claro algo que ya se sabe pero que no se quiere atender.

Sueño n.º 2 - LA DAGA MÁGICA

Estoy entrando en una amplia sala con muchas personas sentadas, podría ser una iglesia o una sala de conferencias, al pasar cerca de una anciana, ésta sufre un desmayo y yo la auxilio, agradecida la anciana me regala algo que parece un juguete, apenas lo miro, no le doy valor pero le agradezco el detalle y me lo guardo en el bolso. La siguiente escena es en una taberna oscura y sucia,

> donde voy a buscar a alguien; hay gente sentada en las mesas y también de pie en la barra; observando a las personas que hay allí tratando de descubrir al amigo que he ido a buscar, veo que muchos tienen como "aliens" o bichos que les están enfermando, algunos están ya casi cadavéricos, entonces veo a mi amigo junto a la barra hablando con otras personas sin enterarse de que tiene uno de esos bichos colgado de su hombro; recuerdo lo que la anciana me ha dado y lo saco del bolso; veo que es una daga que al principio me había parecido de plástico y que ahora veo que es de plata, con incrustaciones de piedras preciosas, una cruz grabada y alguna inscripción; me doy cuenta de que es un instrumento de poder; decido probarla para quitarle el "alien" a mi amigo y funciona; la uso con varias personas más; con alguna otra persona veo que ya es demasiado tarde, que ya no tiene salvación pues los "alien" le han invadido totalmente; empiezo a ver cómo a mí misma se me están enganchando los bichos y mientras me los quito con la daga y viendo que se me van enganchando con gran velocidad y que no doy a basto a quitármelos, pienso que tengo que salir de ese lugar contaminado cuanto antes; salgo corriendo y me despierto.

Es un sueño recordado con mucho detalle y que ha dejado a la persona inquieta; dice que es como si le avisase de un peligro. Antes de entrar a ver los posibles peligros que la soñante piense que pueden amenazarla, le propongo indagar en el elemento mágico del sueño:

1.º Dice que se lo han dado como premio por una buena acción, lo que para ella asegura la bondad del objeto, aunque sea una daga.

2.º Se lo da una anciana que para ella tiene que ver con la sabiduría.

3.º Se lo da en una iglesia o una sala de conferencias, ambos espacios para ella tienen un significado ritual, así como el hecho de que sea ante testigos.

4.º Dibuja el objeto con gran meticulosidad; mientras lo hace dice que experimenta un sentimiento casi religioso.

5.º Al describir los elementos dibujados, les va dotando del simbolismo que para ella tienen: los rubíes, los asocia con la pasión, con la vida; la esmeralda del centro de la empuñadura con el corazón y el amor; la cruz grabada, con el poder sagrado que tiene; la inscripción que no entiende la asocia con la antigüedad de la daga.

6.º Después le llega el turno a la reapropiación de los valores o poderes personales proyectados en el elemento soñado, asumiéndolos como propios, lo que hace emocionándose al hablar de su capacidad de amor, y su espiritualidad.

7.º Le cuesta más asumir su potencial defensivo y agresivo.

8.º Sin embargo le es fácil ubicar con quién y cómo se está sintiendo en peligro en los últimos meses y aceptar que está ampliamente capacitada, ahora "que ha visto sus poderes" para defenderse.

9.º Todavía se hace después un pequeño trabajo para que pueda recoger la proyección de su sabiduría en la anciana, que ahora podemos decir que tiene la función de recordarle sus propias capacidades.

10.º Por último, una vez reafirmada en sí misma, se le pide que se identifique con los "aliens" o bichos enfermizos, trabajando en esta parte proyectada de su personalidad, con la técnica de la silla vacía, sobre cuáles pueden ser los elementos de su personalidad que proyecta como capaces de quitar la vida de otros y de ella misma.

En este sueño el dibujo ayudó a identificar los elementos simbólicos que componían, como en un puzzle, el elemento central de la proyección positiva para, sirviéndonos de la reapropiación de este elemento para autoafirmar a la soñante, poder después elaborar e integrar su "sombra" proyectada en esos otros elementos, "los aliens", que la soñante percibe como mortalmente destructivos.

Sueño n.º 3 - LA EXPLOSIÓN

Estoy en un local en el que hay varias salas que se comunican, yo estoy de pie apoyada como detrás de un arco desde el que veo las tres salas. Hay esculturas mías en diferentes lugares, algunas personas del grupo sentadas por diferentes partes y tú (al terapeuta) entras y te diriges hacia un lugar para sentarte, yo sé que una de las esculturas de piedra está a punto de explotar unos segundos antes de que lo haga justo cuando te sientas; caen pedazos pequeños y polvo por todas partes, también sobre tu cabeza. La sensación al despertarme es de incredulidad, "las esculturas de piedra no explotan".

Aparecen varios elementos que se podrían trabajar en este sueño, al estar en contexto grupal, escogemos:

- los espacios, las salas, dado que la protagonista no cambia de ubicación, es decir la perspectiva no cambia, parece querer indicar una situación.
- las esculturas, hechas por la autora del sueño, indicador de identificación.
- las personas sentadas aquí y allá, el grupo, no parecen tener protagonismo, no actúan ni se mueven, son elementos que configuran la situación.
- la terapeuta, tiene dos acciones, entra y se sienta, y recibe la lluvia de trozos de escultura.
- la protagonista, autora de las esculturas, observadora de la escena, y anticipadora del suceso.

Después de pedirle que configure el espacio del sueño ayudándose de cojines, de colocar a algunos de sus compañeros en las posiciones que ocupaban en el sueño y a otros en el lugar de las esculturas, le pido que escoja a alguien que represente al terapeuta mientras ella narra en presente y primera persona lo que va viendo y sintiendo; se muestra con evidencia entonces que la acción principal ocurre en el triángulo: terapeuta-protagonista-escultura. No aparecen signos que muestren emoción de ningún tipo al explotar la escultura sobre la terapeuta, no obstante le pregunto si le pueden hacer daño, contesta que no, le pregunto también si podría ser algo positivo, y dice que tampoco. Su lenguaje no verbal apoya lo que dice.

Le pido que hable primero desde la escultura que estalla. Resumiendo viene a decir que tiene mucha tensión y que puede estallar en cualquier momento.

Después como terapeuta, dice que no sólo lo ve sino que lo siente, pero que no le asusta.

Hasta ese momento me he mantenido fuera de escena, para facilitarle las proyecciones.

Ahora le pido que se siente frente a mí y me diga lo que ha entendido de su sueño; es un mensaje para mí y para ella misma: que estemos atentas porque puede estallar en cualquier momento. Necesita asegurarse de que estoy pendiente de ella, de que la veo y de que sigo su proceso. Se emociona cuando le confirmo que estoy muy atenta y que no me olvido de ella.

- Este sueño muestra la relación transferencial; la imagen que tiene de su padre era que no la veía mucho y que no estaba capacitado para cuidarla; ahora aunque necesita confirmar que es vista por mí, parece haber depositado la suficiente confianza como para poder permitirse que algo de ella explote sin entrar en pánico. De esto hablamos antes de preguntarle qué piensa que le puede hacer romperse en pedazos. Ve claramente cómo su autoimagen se resquebraja con el proceso que está viviendo en este curso, pero se considera capaz de asumir los cambios y añade, un poco vergonzosa, "sabiendo que estás ahí, por si hace falta".

Sueño n.º 4 - EL ASCENSOR

Es un sueño recurrente que ha tenido ya con anterioridad, y dos veces en el último mes.

Estoy en un ascensor pequeño, antiguo o viejo, no sé, y voy con muchas personas, vamos apretujados. Le doy a un piso, sube y se pasa. Le vuelvo a dar, baja y se pasa. Vuelvo a dar al botón del piso que yo quiero, sube y se vuelve a pasar. Así sigo, sin conseguir que se abran las puertas en mi piso, con la gente pegadísima a mí hasta que me despierto agobiada.

El sueño no muestra elementos visuales ni verbales, es ante todo kinestésico, por lo que le propongo emplear el recurso de la *Performance*; para ello es importante considerar de qué manera modificar el espacio de la escena, ya que nuestras actitudes cambian con él.

1.º Elegir el lugar e iluminación del ascensor, marcar el espacio con cojines de los que hay en la sala y definir el sitio concreto de los botones.

2.º Recrear el ascensor ayudada por personas que se colocan formando un cuadrado mirando hacia el exterior, dos de ellas harán también de puertas.

3.º Escoger de entre el grupo, las personas que van con ella en el ascensor.

4.º Colocarse bajo sus indicaciones dentro del ascensor (se coloca en el medio de otras tres personas con las cabezas casi rozándose).

5.º Recrear corporalmente el movimiento de subir y bajar (lo hacen subiendo los talones un rato y después bajando bruscamente las rodillas en la supuesta frenada).

6.º Repetir varias veces la secuencia: doy al botón, sube, para, abre las puertas, ve que se ha pasado, vuelve a dar al botón, se cierran las puertas, baja, para, se abren las puertas, ve que se ha pasado, etc...

7.º Se va dando cuenta de:

- no sabe cuál es el piso que está marcando
- se le cansan las piernas, las tiene muy tensas
- en ningún momento se le ocurre bajarse del ascensor
- le agobia la proximidad de las otras personas pero no se siente en peligro

8.º Siguen repitiendo la secuencia y después de unas cuantas veces, aprecio en la soñante, tensión de mandíbulas, algunos *tics* en el movimiento de cabeza, y acortamiento de la respiración. De las tres personas que la acompañan en proximidad y movimiento, dos parecen tranquilas y sonrientes y una también parece algo inquieta. Las que forman el ascensor parecen atentas y algo cansadas.

9.º Deja por decisión propia de pulsar el botón del ascensor, con lo que detiene la secuencia.

10.º Le pregunto qué pasa y contesta: *"no voy a seguir pulsando el botón si no sé a qué piso quiero ir"*.

Entiendo que es un "insight" y se lo repito para fijárselo, pero quiero que recoja sus sensaciones corporales antes de que se le difuminen, por lo que le pido que preste atención a cómo está corporalmente y nos lo diga.

- lo que más siente es la tensión en las rodillas, que define como *"entre sujetarse y echar a correr"*.
- la respiración dice que algo agitada *"por el agobio pero no tengo miedo"*.

Su última frase nos lleva a las emociones; le pregunto entonces cómo se siente con las personas que tiene al lado.

- *"eso es lo más curioso, me agobian pero también me dan calorcito; estoy pensando que quizás por eso no salgo del ascensor"*.

11.º Ha llegado el momento de, sin cambiar de posición, mirar a cada persona a los ojos y decir si la podría asociar con alguien de su vida o si le recuerda, o no, a alguien próximo.

– con cara de sorpresa, dirigiéndose a una de ellas, dice: *"me recuerda la mirada sonriente de mi madre, y me agobia";*
– mirando a otra, *"como es más alta siento que me protege, me gusta, pero quizás es un poco demasiado protectora, podría ser mi padre" ;*
– y mirando a la tercera persona *"¡qué mal rollo! es mi hermana mayor, me entiendo bien con ella, pero estoy harta, la tengo hasta en la sopa, siento que me juzga, que está pensando que por qué no salgo del ascensor de una vez"*

12.º "¿Y tú qué le contestas?", le pregunto, a lo que sin dudarlo, muy metida en el papel dice:

"porque os quiero y me da miedo perderos y además no sé cómo me va a ir en otra casa con otra gente".

Después de un rato, añade como recapitulando: *"Sé que quiero irme de casa pero no sé si es el momento".*

Ahí está el dilema que hace que el sueño se repita, está paralizada por una decisión que no termina de tomar, mientras ella misma se está urgiendo a tomarla, quizá antes de tiempo. Parece que la solución que ha encontrado en la elaboración del sueño, ha sido parar la repetición de la secuencia diciendo *"no voy a seguir pulsando el botón si no sé a qué piso quiero ir"* y cuando retomando esa frase le pregunto qué puede significar para ella, dice: *"pues eso, dejarme en paz, hasta que yo sepa lo que quiero".*

Con eso cerramos el trabajo con ella y pasamos a trabajar con algunas de las personas del grupo que han colaborado y que se han identificado con algún aspecto de su rol.

que
estudiar
leer
limpiar

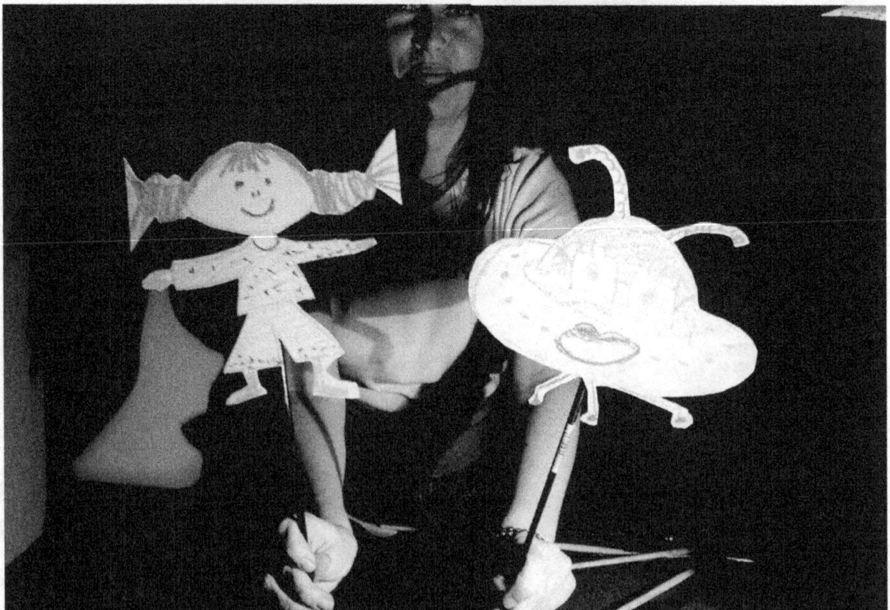

MICRORRELATOS COLOREADOS

INDIGO

"Detrás de las nubes grises
el cielo siempre es azul",
dijo el sabio, anónimo.

Y otro anónimo, quizá también sabio, le contestó:
"Cuando llueve los pájaros no cantan
aunque tras las nubes brille el sol".

Violeta 10

Relativo al Chakra Sahasrara

Hacia un nivel de comprensión mayor

La persona y el grupo

Ejercicios

— Pintura Meditativa Zen

— Meditación en Movimiento: Danza Derviches

— Mandala grupal

❖ CONSIDERACIONES RELATIVAS AL CHAKRA SAHASRARA

El 7.° chakra o chakra corona, se sitúa según culturas en la coronilla o algunos centímetros por encima de ésta, rige el funcionamiento de lo que viene en llamarse mente superior para indicar que no es la actividad pensante habitual sino un tipo de pensamiento unitivo, ligado a un sentimiento de serenidad e integración.

Su cualidad es la espiritualidad entendida como un sentimiento de unidad, de las diferentes instancias personales y de uno con el todo, así como de lo interno y lo externo.

El color con el que se relaciona mas habitualmente es el Violeta luminoso, lo cual tiene un carácter simbólico sobre la propia composición de este color que aúna la calidez del rojo con la frialdad del azul y la luminosidad del blanco que contiene en sí todos los colores del espectro.

Es el chakra en donde se integra la información de todos los chakras, por lo que al configurarse como una unidad de información mayor, puede trascender el nivel anterior.

Para trabajar con este centro es básico tener una buena toma de tierra, estar bien enraizado, para así evitar la disfunción más habitual correspondiente a este chakra: la falta de contacto con la realidad y, en casos mas agudos, la demencia. Con frecuencia se observa otra dificultad al tratar con este centro, la asunción de normas culturales ajenas en el intento de satisfacer la necesidad espiritual que aparece: "hay que ser bueno", "no hay que comer esto o aquello", "sexo sí, sexo no"... y la competición por el método mejor.

El ego, la importancia personal, se manifiesta aquí como un deseo nuevo, revestido de buenas intenciones y con buena prensa, pero tras el cual sigue estando la avidez de conseguir algo, ya sea de tipo espiritual o psíquico, y la dificultad de aceptar lo que hay, es decir, la realidad presente de la persona.

El conflicto en este punto del proceso se concreta en que muchas de las personas que componen el grupo han descubierto gran parte de los comportamientos que les complican la existencia, y comprenden el origen de éstos y cómo han llegado a desarrollar estas formas no funcionales ahora, pero que en su origen sirvieron para satisfacer una necesidad prioritaria, como supervivencia, cariño, o aceptación.

Es el momento en que la persona se plantea ¿y ahora qué?

Normalmente, llegados a este punto, se han ido modificando algunos comportamientos lo que ha traído como respuesta del entorno también algunos cambios; esto por lo general suele ser satisfactorio, la persona se siente más protagonista de su vida y escucha con satisfacción, que algunas personas próximas la vean cambiada.

Pero también es mucho más consciente, vive con más atención lo que le ocurre y lo que hace, por lo que en ocasiones se desespera viendo cómo repite comportamientos que conoce, que no le gustan y que querría ya erradicados. Es, en esos momentos, en los que mas peligro tiene de juzgarse inútil de mejorar, y ser atacada por nuevos impulsos perfeccionistas en seguimiento de nuevos mandatos *introyectados*.

Es un momento difícil del proceso, porque habiendo afinado su capacidad de darse cuenta, observa que está metida en una rueda en la que sus comportamientos se repiten aunque con formas más refinadas y sutiles, pero con el inconveniente de que ahora ya no puede auto-engañarse.

CÓMO ESTÁ EL GRUPO

M. está más expresiva que nunca, cuando se ríe lo hace con todo el cuerpo y sus lágrimas por fin encuentran la salida, en lugar de quedarse enganchadas en su garganta. Cuando comenzó este viaje, hablaba de su trabajo en el campo de la medicina, que le gustaba y le ocupaba muchas horas, de sus dos hijos pequeños, de su marido que viajaba; recuerdo sus ojos dulces y su mandíbula tensa mientras hablaba de todo ello y de la falta de tiempo para sí misma; las tensiones se iban manifestando en pequeños gestos corporales al hablar pero era incapaz de darse cuenta del alto grado de frustración y ansiedad que estaba acumulando. Ahora está en fase de negociaciones permanentes, es trabajoso, pero su salud ha mejorado; es capaz de transmitirnos su enfado cuando cuenta alguna pelea por conseguir una mejora en el reparto de tareas, y observa espantada cómo sigue teniendo la tendencia a olvidarse de su necesidad para no tener que discutir. Está decidida a seguir atenta a este objetivo de observar y pelear por su necesidad, porque como dice "en ello me va la vida" pero no creo que quiera continuar con el proceso de momento, necesita asentar durante un tiempo esta nueva forma de vivir con más escucha interna. Seguramente será una de las personas que se bajen del barco en el próximo puerto, entiendo muy bien que ésa es su necesidad y que es bueno que la escuche, y una parte de mí siente la tristeza de la despedida.

El resto del grupo están muy conscientes de que están a mitad de un proceso, se sienten muy arropados por los compañeros y deseosos de seguir adelante.

El objetivo a trabajar con este chakra, es: **CAMINAR HACIA UN NIVEL DE COMPRENSIÓN MAYOR** y para ello vamos a emplear diferentes técnicas de concentración y meditación, unas estáticas y otras en movimiento, unas silenciosas y otras con música; se trata de ofrecer diferentes opciones para que cada persona encuentre aquella con la que sintoniza mejor.

PINTURA MEDITATIVA - Pintura Zen

Vamos a iniciar una serie de ejercicios que son en sí técnicas de concentración que nos facilitan, a los que estamos educados en la actividad permanente y con un sistema nervioso sobre-estimulado por la vida cotidiana, poder ir ralentizando nuestra actividad física y mental, y disponernos para la verdadera meditación o contemplación.

Vamos a preparar el cuerpo con estiramientos que incluyan la musculatura posterior de las piernas, y la parte alta de la espalda y nuca principalmente, ya que son las zonas que van a estar más implicadas en el trabajo posterior.

A continuación, una música movida, rítmica, que estimule a soltarse, a bailar, a la descarga de tensiones durante quince minutos, al menos, para pasar después a una música más lírica que ayude al cuerpo y a la respiración a encontrar un ritmo más suave. No se trata ya de trabajar intensamente el desbloqueo de alguna zona corporal como hacíamos en otros chakras, sino de, partiendo del estado de agitación habitual acompañar al grupo y a cada persona que lo compone a un estado más relajado, que haga posible acceder al silencio interior.

Bailad de forma libre siguiendo el ritmo de las músicas, y cuando éstas terminen quedaros de pie, quietos y en silencio, con la atención puesta en observar la respiración.

Manteniendo esa actitud nos sentamos situando delante nuestro varias hojas de papel, una piedra de tinta, el recipiente para hacer la tinta, agua y pincel.

Cuando os sentéis intentad mantener el silencio interior, mientras preparáis la tinta; para ello humedeced el recipiente de tanto en tanto y frotad la piedra de tinta; mediante el roce y el contacto con el agua ésta irá soltando la tinta en forma líquida.

La atención en la respiración y el movimiento repetitivo, al tiempo que la atención puesta en la densidad que va alcanzando la tinta, hacen que esta preparación de la tinta sea a la vez la preparación idónea para que la persona entre en un estado de percepción más sutil.

Ya tenemos la tinta y el papel listos, ahora es necesario mantener la atención y no lanzarse con precipitación sobre cualquier objetivo pintable.

Veamos antes cómo estamos a un nivel profundo y tratemos de equilibrarnos; para ello trazaremos unas líneas de izquierda a derecha, siguiendo una pauta repetida, inspirar cargando el pincel de tinta, exhalar deslizando la punta del pincel sobre el papel, con un movimiento del brazo —no de la muñeca— repetimos una y otra vez, observando el paralelo entre los movimientos de nuestra mente y el trazo que queda en el papel; cuando consideramos que es suficiente, repetimos haciéndolo ahora de derecha a izquierda, y después de arriba hacia abajo y de abajo hacia arriba.

Observaremos cómo ha cambiado nuestro estado anímico, después de que con este sencillo ejercicio hemos concentrado la atención visual en la línea que iba surgiendo, al tiempo de realizar un movimiento que involucraba brazo y parte alta del cuerpo y respiración, de una manera unificada.

En Japón estos ejercicios pueden durar meses, siendo seguidos después por el trazado de círculos, líneas quebradas, pinceladas en seco, pintura repetida de objetos sencillos, y así durante años.

No queremos nosotros probar hasta ese punto la paciencia de nuestro grupo, por lo que después de incluir algunos trazos para habituarse a cargar el pincel con diferentes densidades de tinta (más aguada en el fondo del pincel y más densa en la punta) para saber darle volumen a cada pincelada, pasamos a la observación y expresión de un objeto significativo para la persona o el grupo, haciendo que éste sea pintado entrando cada vez más en la esencia del objeto, que es lo mismo que decir nuestra propia esencia.

Se colocan varios objetos esparcidos en el espacio que queda en el centro del grupo, para que desde su posición puedan verlos y escoger concentrar su atención en el que deseen.

Es un ejercicio de visión en atención absoluta, cuyo objetivo último está en sentir aquello que se está pintando, hasta el punto de que desaparezcan las barreras entre lo interior y lo exterior, entre tú y el objeto.

A veces lo que surge de esa concentración es una línea que indica el movimiento del objeto, o puede ser que sea una mancha densa de tinta que señala su volumen, o quizás es el espacio negativo entre las formas el que resulta pintado dejando que del blanco del papel emerja la forma del objeto. Concentraros en vuestra respiración y dejaros fluir con atención y espontaneidad.

En nuestra forma de practicar la pintura meditativa, el cómo resulte es una expresión de libertad; lo importante para llegar a esa libertad y espontaneidad atenta es haber conseguido previamente la conexión con uno mismo y con el objeto, cuanto más profunda sea menos barreras entre ambos.

Recuerda que como en cualquier propuesta de crecimiento personal, cuando se habla de objetivo es para indicar solamente la dirección en la que se camina, y que lo importante, es cada paso del camino.

Quiero contarte la piedra del camino con la que yo me encontré al inicio de esta práctica de pintura zen. Llevaba ya varias semanas en Japón con prácticas de meditación y yoga, por fin iba a conocer a la maestra Rempo con la que iba a hacer estas prácticas meditativas a través de la pintura Zen que había sido el motivo principal de mi viaje.

Durante mucho tiempo me había sentido atraída por este tipo de obras en las que con elementos tan simples como agua, papel y tinta, y con el mínimo de pinceladas, algunos artistas eran capaces de expresar tanto. Había hecho algún trabajo en la facultad estudiando las obras más conocidas e incluso había intentado hacer algunas prácticas por mi cuenta, con lo cual ya no me consideraba novata del todo.

Ya mientras frotaba la piedra de tinta empecé a escuchar mis pensamientos con toda clase de charla, dudas, cuestionamientos, en fin, supongo que en su estado habitual pero como con altavoz; continuaron éstos mientras llenaba papeles y papeles con líneas de derecha a izquierda; era curioso observar cómo cuando juzgaba que una línea me estaba saliendo bien, inmediatamente, antes de terminar el pensamiento la línea se había torcido o había ocurrido algún percance. Por supuesto sabía, en teoría, que no había mal ni bien, pero al parecer mi juez interno no estaba enterado. Aquella noche reflexioné mucho sobre ese aspecto de crítico implacable que hasta entonces no había visto tan de cerca.

Al día siguiente le llegó el turno a las líneas de arriba abajo; el manejo de la tinta, la presión del pincel, la respiración e incluso la postura me resultaban algo más familiares, se podría decir que sentía algo como ¡ya sé de qué va esto! así que inicié la tarea con mayor confianza.

Comencé a inspirar profundamente mientras cargaba mi pincel, apoyaba levemente la punta en la parte superior de la hoja, iba exhalando lentamente, a un ritmo constante, mientras que descendía el pincel hacia la zona inferior del papel, dejando un rastro en ocasiones fino, otras grueso, a veces firme, y otras temblorosas; observaba, trataba de no juzgar, y seguía respirando y llevando el pincel de lo más alejado de mí hacia la zona más próxima.

Empecé a sentirme mal; algo de mareo, opresión en el pecho, ganas de llorar, ¿qué me estaba pasando? "Es absurdo, no me voy a ir" pensaba, "he venido de muy lejos para ahora perdérmelo", y continuaba. Hasta que el malestar se hizo insoportable y la maestra me acompañó fuera, me escuchó, me comprendió, y me explicó con un gesto la similitud con el movimiento del harakiri, no había dramatismo en su explicación sino ayuda para el entendimiento, entendí el paralelo en la actitud de entrega y en el camino de la herramienta que se usa para ir hacia el interior, en el caso del samurai podía ser la daga, en el mío podía ser el pincel, menos sangriento pero, quizás, no menos doloroso.

He querido compartir este recuerdo para ilustrar mi opinión de que aunque es una experiencia que se puede hacer en solitario, al menos al principio, o de tanto en tanto, conviene tener alguien a quien acudir para que nos apoye u oriente, para que nos ayude a ver y no pasar por alto, las piedras del camino.

CÓMO ESTÁ EL GRUPO

Cada componente del grupo ha adquirido ya bastante capacidad de introspección, han superado muchas resistencias al contacto real entre ellos, es decir, se relacionan con mayor autenticidad, y en estos últimos módulos se trata de que se conecten con una visión más holística, que incluya lo personal y lo trascienda; que sin perder la atención sobre sí mismos sean capaces de sentir profundamente al otro; que sintiéndose parte del todo y del grupo, también se sientan libres de continuar en él o de separarse atendiendo a su necesidad interna, porque estamos terminando esta etapa del viaje y necesitan decidir con responsabilidad, si es su momento de retirarse o si continuar con una implicación aún mayor, para formarse profesionalmente en el acompañamiento de otras personas.

Por ello la siguiente propuesta es una meditación que tiene que ver a la vez con lo individual, lo grupal, y lo trascendente.

DANZA DERVICHE

Muchas tradiciones espirituales nos enseñan que la meditación en quietud y silencio es buena para el espíritu y necesaria para nuestro equilibrio, pero que debe servirnos como práctica o entrenamiento para ser capaces de cultivar esa actitud meditativa, de atención y desapego en la vida cotidiana.

En esta línea, las danzas derviches, comúnmente llamadas *"Baile Sufí"* que desde hace mucho tiempo practican *"los giróvagos"* pretenden a través de la rotación conectar la persona con la Unidad. A través de la danza, el *sufí* realiza la unión del espíritu con la materia, tomando la luz de los cielos, del espíritu, y con la atención y el punto de equilibrio en el corazón, derrama esta luz en todas las direcciones del espacio. Uniendo la vertical y la horizontal en el centro de su corazón mientras el universo gira sin cesar.

Voy a ir dando paso a paso las consignas para una primera práctica de este tipo de meditación en movimiento. Es posible que algún detalle varíe de cómo lo enseñan unas u otras personas, pero hasta donde yo sé en la esencia todas coinciden: Es importante centrar la atención en el corazón que es el lugar donde se reúnen las energías; el pie izquierdo nunca abandona el suelo porque el trabajo del verdadero derviche se realiza en la tierra; la mano derecha se queda elevada palma arriba y la izquierda hacia abajo, para que al girar seas un canal que recibe la luz de Dios con la mano derecha y la transmite a la tierra por la mano izquierda.

Anclamos muy bien el pie izquierdo, es nuestra conexión con la tierra que en ningún momento podemos perder; mientras, con el otro pie vamos dando pasos, al norte, al oeste, al sur y al este, en las cuatro direcciones. Esta es una de las claves, anclar bien este movimiento primero. Encontrar cuál es el tamaño del paso más adecuado para el equilibrio de cada uno.

En ningún momento se cierran los ojos, la mirada se dirige desenfocada a la palma de la mano derecha, sentid la música dentro del cuerpo, a veces, gira más despacio, a veces más deprisa, quizás haya momentos en los que alguna persona tendrá que atravesar el miedo a caerse o a marearse, si necesitáis parar no lo hagáis de golpe, ir poco a poco parando y después os agacháis y ponéis la cabeza en el suelo, pero si podéis, os recomiendo intentar atravesar el miedo, son solo resistencias; la siguiente cuestión, que el brazo derecho con la palma hacia arriba y el izquierdo con la palma hacia abajo, formen una onda, sin ángulos, relajados los hombros, brazos y muñecas; habrá momentos en los que os apetezca subir o bajar algo la altura de los brazos; mantened la idea de dar y recibir; no olvidéis la respiración.

Centraros en la respiración todo el tiempo. El trasfondo de esta danza es la importancia de estar bien anclados con el suelo, bien conectados con el cielo y dispuestos a recibir y a dar, y con la mirada abierta y relajada, para estar en contacto con el exterior sin que éste nos atrape.

Mientras se realiza la práctica con músicas más lentas y más rápidas, voy intercalando algunas frases, de recordatorio de la respiración, del anclaje y algunas otras como sugerencia.

Fijaros que somos como planetas, cada uno importante en sí mismo, pero a la vez todos nos movemos a un ritmo. No levantéis el pie izquierdo del suelo, y seguid la música.

Manteneros anclados en el suelo, ya sea el movimiento lento o rápido, manteneros, como la actitud, impecable, en el presente, disfrutando de lo que hay.

Sentid que esto es una forma de entrar en contacto con la danza del universo, la danza del cosmos, todos somos parte de esa danza. Más allá de las emociones, de los miedos, de cualquier cosa, está esa conexión, ese juego de dar y recibir, del que todos formamos parte.

Después de un tiempo, la música va siendo más lenta, van parando, les indico que crucen las manos sobre el pecho, que mantengan los pies bien apoyados, que respiren naturalmente, y que si lo desean pueden cerrar los ojos.

Las pinturas que realizan después para recoger esta experiencia, son en su mayoría Mandalas, muestran claramente la conexión interna y externa que han tenido, mientras algunos de sus comentarios lo subrayan y otros ponen el énfasis en los momentos de miedo o en las tensiones corporales percibidas.

I., comparte con el grupo: *"En un momento me vino el recuerdo de algunos episodios familiares, me empecé a marear y me entraron ganas de vomitar, pero continué y se me pasó, y después he llegado a un punto que sentía como que estaba quieta y que era lo de alrededor lo que se movía, como muy anclada, y todo dando vueltas, incluso ralentizándome un poco seguía teniendo la sensación de girar rápido; he llegado a estar super anclada y eso es nuevo para mí".*

C. dice: *"Mientras giraba me he dado cuenta de que me distraía controlando cómo estaban los demás, eso me ha recordado una actitud con mis hijos que me gustaría evitar; otra cosa que me ha llamado la atención y que veo que tiene que ver con mi vida, es que me sentía muy nervioso con los ritmos más lentos, pero cuando la música era rápida me era más fácil, era como dejarme llevar ".*

L., resuena con este comentario: *"yo metí el acelerador hasta el mareo y el vómito. En el lento, no me resultaba fluido. La rapidez me invadía. A veces si alguien chocaba conmigo, me enfadaba".* Alguien sonriendo le pregunta si siempre eran los demás los que chocaban con ella, al principio no lo entiende, luego se ríe y se queda pensativa.

E. cuenta como para ella chocar con alguien o con algo no la distraía, su experiencia ha tenido que ver, dice emocionada, con el fluir, con vivir el presente, dice: *"la idea que me venía era, vivir es fácil".* Hay otras voces que resuenan con esta experiencia, hablando de apertura, del no esfuerzo, de alegría sintiéndose y viendo a los demás.

La experiencia ha sido importante para todos y conviene recogerla plásticamente ya que tiene que ver con la conexión interna que han alcanzado en esta etapa de su formación, y también con cómo se relacionan con el grupo y la vida en este momento, por lo que sacar una imagen visual puede ser una buena forma de cerrar el trabajo de este ciclo.

MANDALAS

La propuesta va a ser un *Mandala* colectivo, pero antes vale la pena mencionar lo más significativo con respecto a los *Mandalas* y las diferentes formas de trabajar con ellos.

— Se puede trabajar con pinturas del tipo Mandala, como diagramas que representan e integran diferentes aspectos de la vida de la persona.

- También se hacen pinturas de este tipo sobre superficies previamente dibujadas dispuestas para colorear, con una intención armonizadora.
- Se realizan a veces, sobre círculos parcelados mediante verticales y horizontales que faciliten después la lectura como proyección de diferentes partes corporales o psíquicas.

En nuestro caso, durante los ejercicios relativos al chakra Indigo, se han realizado este tipo de pinturas como representación del individuo y su universo, en el sentido de relaciones, áreas de interés y actuación, etc., tratando de aunar también consciente e inconsciente.

Recordemos que históricamente servía de símbolo para aclarar filosóficamente la naturaleza de la divinidad o representarla en una forma visible. Más recientemente Jung, refiriéndose a los que las personas ven en sueños o visiones, o a los que desarrollan por *"imaginación activa"*, dice que expresan, una actitud que no podemos menos que llamar "religiosa", y subraya *"Un mandala moderno es una confesión involuntaria de un particular estado espiritual"*.

A esta forma más profunda de realización de un *Mandala* es a la que estamos acudiendo aquí; la preparación para ello no ha sido mediante una visualización o imaginación activa sino que se ha realizado mediante una meditación en movimiento *—la danza derviche—* que es ya de por sí un *mandala* en movimiento. Así, de una manera espontánea, las pinturas que han surgido eran en muchos casos de tipo *mandalas*.

Ahora la propuesta es más directiva, en atención a que se realice un cierre grupal, en el que cada uno contemple la imagen del "sí mismo", con el significado junguiano de unidad consciente-inconsciente, en relación con el grupo, unido y diferenciado a un tiempo, para que pueda producirse una despedida consciente con las personas que dejarán el grupo al finalizar esta fase.

Para ello montamos entre todos un papel cuadrado de dos metros y medio de lado en el centro de la sala, les pido que dibujen y recorten un gran círculo,

> *Pintad la experiencia que habéis tenido mientras girábais, cómo os sentíais en relación con la tierra y el espacio, y también con respecto a los que giraban a vuestro alrededor.*

Hay un silencio profundo mientras que lo hacen, me siento emocionada viviendo este momento en esta pequeña parcela del universo.

El resultado se parece mucho a un mapa del cielo, con planetas, estrellas y constelaciones, con luces que irradian de algunos y llegan a otros.

Sé que a algunos de estos planetas no voy a volver a verles y les he llegado a conocer tanto, que es imposible no amarles.

Van terminando; algunos se suenan, algunas se limpian alguna lágrima, otros sonríen. La propuesta del trabajo termina invitándoles a que se repartan el universo grupal; llegaron con sus equipajes, se enriquecieron con la luz de los demás planetas mientras irradiaban su luz propia, y ahora, el viaje continúa; algunos se quedarán algún tiempo, otros se van, pero todos se llevan su planeta más colorido.

En otras ocasiones, ante una propuesta de reparto de algo colectivo, han surgido rápidamente los que no quieren quedarse sin esa parte por la que se sienten atraídos; esta vez sin palabras todos se quedan mirando "su universo grupal" sin moverse. Parece que no hubiera prisas en escoger y llevarse sus beneficios. Empiezan a aparecer vocecillas de "preferiría dejarlo como está" "no quiero romperlo" hasta que deciden dejarlo instalado en una pared del centro "al menos durante un tiempo".

❖ CONCLUSIONES

Una vez más lo simbólico expresado en el trabajo plástico, está diciendo no solo cómo está el grupo, sino también sus intenciones. Salvo alguna persona que tiene que dejar el grupo por circunstancias personales, el resto no sólo va a continuar sino que han entendido profundamente el valor del trabajo en equipo, y sienten su individualidad pero también su entidad grupal. Están en un momento estupendo para continuar con su proceso a través del ciclo superior, confrontándose con todas las dificultades que supone el entrenamiento en acompañar a otros viendo al tiempo, el funcionamiento de los mecanismos propios en acción; la importancia personal a veces dolida, pero con el enorme apoyo que supone sentirse acompañada y querida por el grupo, aceptado, aceptándose y aceptando.

Pero ya que estamos cerrando procesos, me gustaría apuntar un poco más lejos, aunque solo sea para abrir interrogantes, para seguir haciéndonos preguntas, para seguir vivos.

Ramesh Balsekar, maestro de la enseñanza Advaita, distingue entre la mente funcional y la mente pensante. La mente funcional es la que respondiendo a la predisposición genética y los condicionamientos adquiridos, determina cómo actuamos en cada momento, es útil y necesaria. La mente pensante es la que reclama la autoría, y dice "Yo pensé eso; Yo decidí eso", es la que se identifica y la causa principal de sufrimiento; para la enseñanza Advaita, es prescindible. Estamos realizando un trabajo con el ego, así que, hasta aquí, estudiamos y aceptamos esas dos subespecies de mente, pero ¿qué es lo que hace que estas personas y no otras, decidan en este momento seguir girando juntas? ¿qué es lo que hace que surja la necesidad, la voluntad y el impulso de que este libro sea escrito, impreso, e incluso leído?

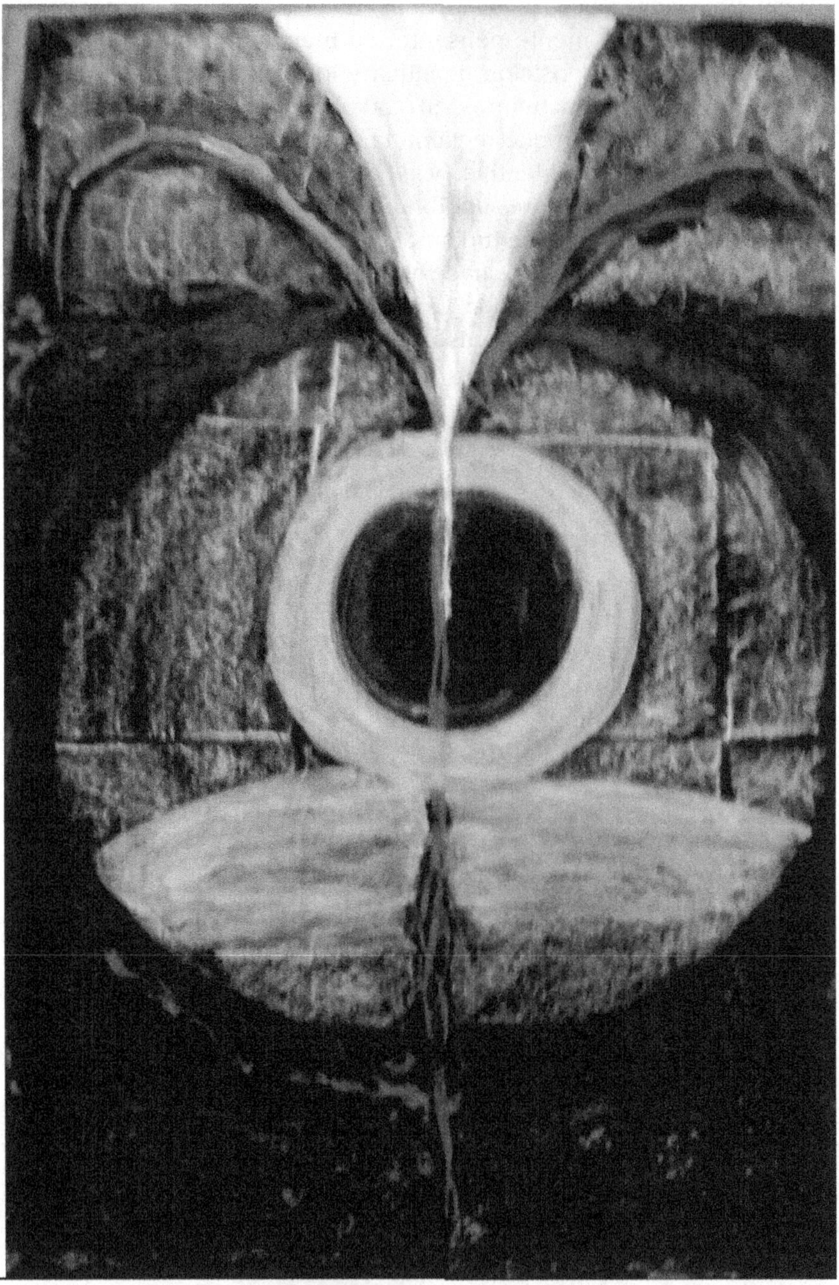

MICRORRELATOS COLOREADOS

VIOLETA

Para alcanzar la maestría en tu arte, dijo el maestro, debes trascender la técnica de manera que la acción surja directamente del inconsciente.

El discípulo, pintor de marinas, al cabo de muchas respiraciones dio una última pincelada y se dejó arrastrar por las olas.

Conclusiones 11

Es necesario recordar lo que se decía en la Introducción respecto a los tres ejes de esta pedagogía, que actúan en paralelo: el eje vertical que sigue el recorrido de los chakras, y que se ha podido ver a lo largo del libro como actúa sobre los diferentes aspectos: físico, mental, emocional y espiritual; el eje horizontal que tiene que ver con lo relacional, que está presente en todo momento y que es motivo frecuente de intervenciones terapéuticas y que no tenía cabida en este libro; y el eje en profundidad que es el efecto que el trabajo va ejerciendo sobre la estructura egoica y que se va manifestando en cambios de actitud más que en verbalizaciones.

Todo lo que constituye el trabajo con el eje horizontal, que no ha sido posible recoger en este libro, está basado en la observación de los juegos de transferencias, en la toma de conciencia de los roles en el comportamiento social y en el análisis de las dinámicas de grupos que se explicitan a través de las acciones artísticas. Pero como dicen en los cuentos, esto es otra historia.

Espero haber conseguido transmitir que el desarrollo personal a través del Arteterapia Humanista lo entiendo como un proceso. Dicho proceso se lleva a cabo ayudado por lo que en términos profesionales se denomina "*setting*" y que yo acostumbro a identificar con la idea de laboratorio, en el sentido de que es investigación, en un ámbito de seguridad, facilitada por la presencia del arteterapeuta.

— Investigación, porque se hacen descubrimientos, se pueden probar comportamientos nuevos, y requiere del componente de la curiosidad.

— Ámbito de seguridad, porque las normas de funcionamiento, como la confidencialidad, la evitación de juicios, y el propósito compartido, generan la confianza imprescindible.

— La presencia del arteterapeuta, sobre esto quiero apuntar algunas cosas.

Cuando hablo de presencia, hablo de presencia plena, y eso ni se improvisa ni se aprende en un cursillo; la presencia es estar en contacto con el

Sí Mismo, consciente-inconsciente, y con apertura hacia el otro, que se manifiesta en algunos comportamientos como:

- Empatízar para comprender su punto de vista, y fascinarse con su historia.
- No juzgar en positivo, no juzgar en negativo, no juzgar, cada opinión y cada obra es digna de ser tenida en cuenta.
- Cultivar la capacidad amorosa para apoyar o confrontar cuando es necesario.
- Respetar su ritmo, sin dejarse llevar por la impaciencia propia (del arteterapeuta).
- Ayudarle a que sea quien descubra, a que sea el protagonista de su vida, no sustituirle.
- Sin interpretar, siendo capaz de renunciar a la posición "del que sabe" que deja al cliente en posición de depencia.
- Recordar que cualquier imagen tiene múltiples significados y en nuestro caso, el mejor es el que le da su creador.
- Atender a lo que cuenta y a lo que no cuenta.
- Y algunas cosas más.

Winnicott (1971) redefinió la terapia como *"un juego entre dos, terapeuta y paciente"* pero no olvidemos que el juego que se establece entre el terapeuta y el cliente, requiere que ambos sepan jugar; esa es una capacidad a recuperar y a potenciar.

La acción creativa en la *triangularidad* cliente-obra-arteterapeuta, determina un espacio transferencial terapéutico, gracias a ese espacio intermedio *(transicional*) que constituye la obra, que en sí misma se vuelve "contenedora" de significados y de relación; por lo que es sumamente importante que el arteterapeuta sepa respetarla y no cargarla con sus propios significados interpretándola, al tiempo que tiene que estar disponible para escuchar y devolver al cliente, todo cuanto la acción tenga que ver con lo relacional.

Resumiendo, acompañar a las personas en su proceso de indagación personal y del significado de su simbología, requiere haber hecho uno mismo, previamente, ese recorrido, para poder comprender las dificultades, las trampas, y los estímulos. Formarse no es solamente el conocimiento de técnicas y conceptos, sino la profundización en el propio trabajo personal.

Bibliografía

Tema Arteterapia

Arnheim, R. (1969) *Visual Thinking*, Berkeley, univ. California Press.

Allen, P. B. (1997) *Arte-Terapia. Guía de autodescubrimiento a través del arte y la creatividad*. Ed. Gaia, Madrid.

Cirlot, J.E. (1949) *Diccionario de los ismos*. Ed. Argos, B. Aires.

Dalley, T. (1987) *El arte como terapia*. colecc. Biblioteca de Psicología, ed. Herder, Barc.

Dominguez-Toscano, P.(2004) *Arteterapia. Principios y ámbitos de aplicación*, Junta de Andalucía.

Gutiérrez, E.(1999) *Arteterapia con orientación gestáltica*. tesis doctoral Univ. Complutense, Madrid.

Gombrich, E.(1963) *Meditaciones sobre un caballo de juguete*. Seix Barral, Barcelona.

Klein, J.P. (2004) *L'art-térapie* P.U.F. París.

Liebmann, M., (1942) *Art therapy for groups*, Routledge, London.

López Fdez.M. (coord) (2006) *Aplicaciones del arte en la integración social: arte,*.ed. Fundamentos, Madrid.

Lowenfeld (1947), *Desarrollo de la capacidad creadora*. Kapelusz, Buenos Aires.

Martínez Díez,N. y López Fdez. M. (2006) *Arteterapia. Conocimiento interior a través de la expresión artística* ed. Tutor, Madrid.

Pain, S. y Jarreau, G.,(1995) *Una psicoterapia por el arte. Teoría y Técnica*. Nueva Visión. Buenos Aires.

Sudres, J. L., y otros (2003), *Humeurs et pratiques d´art thérapie*. L`Harmattan, París.

Widmer,K. (2006) *Pintura-Terapia Gestáltica*. Mandala ed.

Winnicott, D. (1971) *Realidad y Juego*, Gedisa, Barc.

Tema Gestalt

Naranjo, C. (1990) *La vieja y novísima Gestalt*. C.Vientos, Chile.

Naranjo, C. (1993) *Gestalt sin fronteras* Era Naciente, B. Aires.

Oaklander, V. (2003) *Ventanas a nuestros niños. Terapia gestáltica para niños y adolescentes*. ed. Cuatro Vientos, Sgo. Chile.

Peñarrubia, F. (1998)*Terapia Gestalt. La vía del vacío fértil* ed. Alianza, Madrid.
Perls,F.(1976) *El enfoque gestáltico. Testimonios de terapia.* Ed. Cuatro Vientos, Chile.
Perls, F.S., Hefferline, R., Goodman, P. (2002) *Terapia Gestalt: Excitación y Crecimiento de la personalidad humana*, Madrid-Ferrol, CTP.
Salama, H y Castanedo, C (1991) *Manual de Psicodiagnóstico, Intervención y Supervisión* El Manuel Moderno, México.
Yontef, G.M. 1995 *Proceso y Diálogo en Terapia Gestalt* Cuatro Vientos, Santiago de Chile.
Zinker,J. (1999) *El proceso creativo en la terapia gestáltica*, Ed. Paidos México.

Tema Psicoterapia Humanista

Jung, C. G. (1994) *Arquetipos e Inconsciente colectivo*, ed. Paidós, Barc.
Lewin, K. (1974) *Teorías contemporáneas del aprendizaje*, ed. Paidós, Buenos Aires.
Maslow, A. (1987) *La personalidad creadora* Kairós, Barc.
Reich,W. (1995) *La función del orgasmo*. Paidos. Barc.
Rogers,C.R. (2002) *El proceso de convertirse en persona*, Paidós, Barc.

Tema Espiritualidad

Balsekar, R. S. (1982) *El Buscador es lo Buscado*, ed. Gaia, Madrid.
Dürckheim, K. G. (1992) *Hara*, ed. Mansajero, Bilbao.
Jung,C.G.(1949 1ª ed) (1994) *Psicología y Religión* - Prólogo, supervisión y notas de Enrique Butelman ed. Paidós, Barc.
Liquorman, W. (2006) *No le des vueltas.* La trompa del elefante. Madrid.

Tema Chakras

Brennan, B. A. (1990) *Manos que curan* Edic. Martinez Roca, Barc.
Katzeff, M.(1988) *Psicología de los chakras.* Mandala ed.
Leadbeater, C. W. (1991) *Los chakras* ed. Humanitas, Barc.
Merleau-Ponty, M. (1985) *Fenomenología de la Percepción*, Planeta-Agostini, Barc. 1985.

Rojo

Levy,N. (1999) *La sabiduría de las emociones* ed. Sudamericana, Barc.

Lowen, A. (1995) *La traición al cuerpo. Análisis Bioenergético* ed. Era Naciente, B. Aires.

Stevens J. (1976) *El darse cuenta* Ed. Cuatro Vientos Chile.

Naranja

Cadoux, B.(1999) *Écritures de la psychose* Aubier, Paris.

Pease, Allan *"El lenguaje del cuerpo"(*1989) Paidós, Barc.

Schnake, A. (1995) *Los diálogos del cuerpo. Un enfoque holístico de la salud y la enfermedad* C.Vientos, Chile.

Vaysse, J.(1997) *La danse-thérapie.* Desclée de Brower, Paris.

Amarillo

Castanedo, C. (1986): *Terapia Gestalt: Enfoque del aquí y ahora*. Pág. 109-116. Herder. Barc.

Castaneda,C. (1994) *El conocimiento silencioso*, Gaia ed. Madrid.

Enright, J.B: "Introducción a las técnicas guestálticas" en Fagan, J. y Sepherd, I. (1989): *Teoría y técnica de la psicoterapia guestáltica*. Amorrortu edit. B. Aires.

Landy, R. (1983) "The use of distanncing in Drama-Therapy. En *The Arts in Psychotherapy*, vol 10.

Lecoq. J.(2003) *"El cuerpo poético. Una pedagogía de la creación teatral"* Alba edit. Barc.

Salama P., H. y Villarreal B., R. (1988): *El enfoque guestalt. Una psicoterapia humanista* Ed.El Manual Moderno, México.

Sheleen L. (1983) *Théâtre pour devenir..autre* l´¨Epi, París, 1983.

Verde

Fagan y Sheperd (2001) *Teoría y técnica de psicoterapia gestáltica*. Ed. Amorroetu, Buenos Aires.

Kepner, James I. (1992.) *Proceso corporal. Un enfoque Gestalt para el trabajo corporal en psicoterapia.* Manual moderno.

Reich, Wilhelm (1995) *La función del orgasmo.* ed. Paidós, Barcelona.

Celeste

Betés de Toro, M.y otros, *Fundamentos de Musicoterapia*. Ed. Morata.
Benenzon, R. (1997) *Sonido, Comunicación, Terapia*, Amaru edic.
Manchado, M. L. (2006) *Musicoterapia gestáltica*. Mandala ed., Madrid.
Wilfart, S.(1999) *Encuentra tu propia voz. La voz como instrumento para tu desarrollo personal. Ed.Urano, Barc.*

Indigo

Duflot (1992) *Des marionettes pour le dire* Hommes et perspectives. París.
Jung, C. G. (1955) *El secreto de la flor de oro*. Buenos Aires, Paidós.
Martorell J.L.(2000) *El guión de vida* Ed. Descléé De Brouwer colección Serendipity, Bilbao.
Moreno J. L. (1984) orig.1923 *Théâtre dela spontanéité*. Épi, París.
Watts,A. (1992*) El Gran Mandala*. Kairós, Barcelona.

Violeta

Cheng, Fr. (1995) *Vacío y Plenitud* ed. Siruela.
Franck, F. 1973 *The Zen of Seeing. Seeing/Drawing as Meditation*.ed. Vintage Books - N. York.
Watts, A. (1976) *El camino del Tao*, ed. Kairós.

Nota biográfica de la autora

ELVIRA GUTIÉRREZ RODRÍGUEZ

Doctora en Bellas Artes por la Universidad Complutense de Madrid. Formada en terapia Gestalt, en psicoterapia integrativa, y en terapias corporales y del color. Ha colaborado con las Universidades Autónoma y Complutense de Madrid, y con la Fundación Red Apoyo para la Inserción Sociolaboral. Es miembro titular, docente y supervisora de la Asociación Española de Terapia Gestalt y está reconocida como psicoterapeuta por la F.E.A.P. Directora del Centro de Arteterapia y Terapia Gestalt de Madrid, coordina la formación en Arteterapia Humanista desde 1997. Presidenta de la Asociación Arteterapia Gestáltica Humanista. Autora de la primera tesis doctoral sobre Arteterapia en España "Arteterapia con orientación gestáltica" (1999) y de la tesina de la Asociación Española de Terapia Gestalt "Arteterapia. El color y los centros de energía" (1998).

ELVIRA GUTIÉRREZ RODRÍGUEZ
Directora Centro Arteterapia y Terapia Gestalt
elviragr@inicia.es

FORMACIÓN ARTETERAPIA HUMANISTA
www.arteterapiahumanista.com
informa@arteterapiahumanista.com

OTROS TÍTULOS DE LA COLECCIÓN

Libros:

Pintura-Terapia Gestáltica – Imágenes del alma
Katharina Widmer

Musicoterapia Gestáltica – Proceso sonírico
Marisa Manchado Torres

Pequeña Antología de Salomo Friedlaender
Fragmentos de S. Friedlaender y artículo de L. Frambach

El Berlín de Perls:
el espíritu vanguardista en el arte y la terapia Gestalt
Annie Chevreux

Meditación Vipassana y Gestalt
Dhiravamsa entrevistado por Juan José Díaz

Ternura y Agresividad – Carácter: Gestalt, Bioenergética y Eneagrama
Juan José Albert Gutiérrez

El Espíritu Dionisíaco de Fritz Perls y de la Terapia Gestalt
Emilio Blázquez

DVD:

Autoasistencia Psicológica: De la Exigencia a la excelencia
Norberto Levy

Revista:

Estudios de Gestalt: *Nº 1 / Nº 2*
Varios autores
(Para consultar artículos de cada número: en la página siguiente)

PRÓXIMOS TÍTULOS DE LA COLECCIÓN

Holismo y evolución
Jan C. Smuts

Revista

Estudios de GESTALT

Nº 1

ESTUDIOS

- El proceso del terapeuta gestáltico | ***Antonia Canyelles Socias***
- La ansiedad según la teoría de la Terapia Gestalt y su aplicación clínica | ***Esperanza Miravalles González***
- La gestión de la agresividad en la teoría de la Terapia Gestalt | ***Mª Cristina Rodríguez Inyesto***
- La seducción dentro de la Terapia Gestalt | ***José Julián Gómez López***
- La vergüenza | ***Carlos Cervera López***
- La equivocación, la culpa y el perdón | ***Amor Hernández Fernández***
- Sé un payaso... Sé tú mismo. Clown & Gestalt | ***Néstor Muzo Fredes***
- Arteterapia gestáltica: Herramientas artísticas en el trabajo de polaridades | ***Carmen Pardo Martínez***
- Unas gafas gestálticas para mirar el sistema floral del Dr. Bach | ***Rosa Medinilla Segura***
- El canto de las sirenas. Pelando la cebolla de las drogo-dependencias | ***Carmen Benítez e Inmaculada Casas***
- Aportaciones de la terapia Gestalt al tratamiento del Síndrome de Burnout (quemarse por el trabajo) en los profesionales de la ayuda | ***María Dolores Baena Sánchez***

Nº 2

ENSAYO

- Ante la intemperie existencial: ¿Vertical u horizontal?
  | ***Francisco Huneeus***

ESTUDIOS

- Aproximación a la terapia infantil gestáltica | ***Mercedes Mendikute***
- El uso de la Fantasía Dirigida: aplicaciones en niños y adolescentes
  | ***Yael Peña***
- Maltratador y Cuidador | ***Ágata Asensi Gonzalo***
- Las pérdidas y sus duelos en la vida y en la terapia | ***Jesús Pinedo***
- Música y Gestalt | ***Emilio Blázquez***
- Enteógenes y Gestalt: Un trabajo desde la consciencia
  | ***Raúl Martínez Ibars***
- El proceso de creatividad durante la formación en Gestalt
  | ***Maribel Gómez Sanjuán***
- Placer y Terapia Gestalt | ***Javier Egurzegui Lorza***

CASO

- Arteterapia y adecuación a las circunstancias | ***Elvira Gutiérrez***

Nº 3 (en preparación)

ESTUDIOS

- Desarrollo de la Conciencia Global y de la Responsabilidad Social Corporativa genuina. Una perspectiva Gestalt | ***Eugenio Moliní***

- Yo juego ¿y tú? (método fácil de lectoescritura... la Gestalt también juega) | ***Pepa Campos***

- Aplicación del Enfoque Gestáltico a la intervención en organizaciones | ***Aracelli Déniz Benítez***

- Reconocer la Ternura | ***Oskar Ignacio Ekai Pagoto***

- Terapia Gestalt: una guía de trabajo | ***Francisco Sánchez Gavete***

- Miradas hacia la vida – Ganar al perder | ***María Antonia Plaxats / Cristina Millán***

ARTÍCULOS

- Terapia gestalt: Una perspectiva libertaria de la psicología social de la infelicidad | ***Stephan Blankertz***

- Similitudes entre el teatro off-off-broadway y las consultorías de empresas. Manifestaciones del postmodernismo – una perspectiva social gestáltica | ***Raymond Saner***

GESTALT Y...

- Autoasistencia Psicológica y Enfoque Gestáltico | ***Norberto Levy / Aída Bello Canto***

- La PNL y yo | ***Francisco Huneeus***

- Dimensión transpersonal de la Terapia Gestalt | ***Juan José Díaz***

CASO

- Caso clínico | ***Juan José Albert***

Made in the USA
Monee, IL
18 October 2026

66048462R00103